ENVIRONMENTAL APPLICATIONS OF
NUCLEIC ACID AMPLIFICATION TECHNIQUES

HOW TO ORDER THIS BOOK

BY PHONE: 800-233-9936 or 717-291-5609, 8AM–5PM Eastern Time

BY FAX: 717-295-4538

BY MAIL: Order Department
Technomic Publishing Company, Inc.
851 New Holland Avenue, Box 3535
Lancaster, PA 17604, U.S.A.

BY CREDIT CARD: American Express, VISA, MasterCard

BY WWW SITE: http://www.techpub.com

ENVIRONMENTAL APPLICATIONS OF NUCLEIC ACID AMPLIFICATION TECHNIQUES

Edited by

GARY A. TORANZOS, Ph.D.
Department of Biology
The University of Puerto Rico
Rio Piedras, Puerto Rico

TECHNOMIC
PUBLISHING CO., INC.
LANCASTER · BASEL

Environmental Applications of Nucleic Acid Amplification Techniques

a TECHNOMIC® publication

Published in the Western Hemisphere by
Technomic Publishing Company, Inc.
851 New Holland Avenue, Box 3535
Lancaster, Pennsylvania 17604 U.S.A.

Distributed in the Rest of the World by
Technomic Publishing AG
Missionsstrasse 44
CH-4055 Basel, Switzerland

Printed in the United States of America
10 9 8 7 6 5 4 3 2 1

Main entry under title:
 Environmental Applications of Nucleic Acid Amplification Techniques

A Technomic Publishing Company book
Bibliography: p.
Includes index p. 211

Library of Congress Catalog Card No. 96-60907
ISBN No. 1-56676-408-4

To Alberto and Emma for their love, complete trust and for teaching me the real meaning of:

"Ama Sua, Ama Llulla, Ama Khella"

Madre, el dolor de tu ausencia se esfuma con el recuerdo de tu sonrisa . . .

Table of Contents

Preface

FROM sewage to fossils, microorganisms are present in every nook and cranny. Their presence, forever felt, was seen for the first time by Anton van Leeuwenhoek centuries ago. Eventually Robert Koch gave us the first glimpse of the microorganisms that could be grown in the laboratory. Over 100 years later the techniques for isolating microorganisms have changed little. We have come to realize that most of the microorganisms in nature cannot be grown in vitro. Additionally, when in pure culture, microorganisms behave differently than they do in nature. Although the study of microorganisms in pure culture allows for autecological studies, and is necessary when dealing with pathogens, synecological studies often require direct analysis of microorganisms in the environment.

Antibody techniques have allowed us to study microorganisms *in situ*. However, until recently all methodology lacked the sensitivity necessary for environmental work where microorganisms are in most cases present at very low concentrations or where microbial ecosystems contain a myriad of different organisms. Gene probes have been used successfully for a variety of samples, but this method was still not sensitive enough. The next logical step was the application of the recently developed DNA amplification technique known as the polymerase chain reaction, or PCR. Since then, many laboratories around the world have adopted PCR for environmental work.

If molecular biology has a central dogma, then the polymerase chain reaction has become its catechism. The term invokes a series of infallible experiments. Scientific lore tells of obtaining enormous amounts of PCR products from insignificant original concentrations. The term PCR has even been introduced to the lay public in movies and overblown media events. The analogy used to explain PCR to the uninitiated is that of the needle in a haystack—and one doesn't just look for the needle but amplifies the number of needles to end up with a pile larger than the haystack. It is easy, then, to assume that PCR is infallible.

Researchers in environmental biology repeatedly ask why target DNA in a particular sample does not amplify well by the PCR. Many a graduate student has lost sleep over the repetitious task of standardizing an amplification procedure, only to find that the technique has to be restandardized for a second set of samples. In many cases where amplification does not occur, entire stocks of materials are thrown out and replaced. It does not take long to realize that amplifications, because of their immensely complex nature, are rarely trouble-free when working with environmental samples. One of the most common problems is the presence of inhibitory compounds, such as humic acids and heavy metals. Under some circumstances, the extreme efficiency of PCR can also be a problem, since carryover of even one target molecule can give a false positive reaction.

A lot of literature exists on the use of the polymerase chain reaction. However, relatively little work has been done on environmental samples as compared to clinical samples or pure bacterial cultures. Samples obtained from soils, water, and air are enormously complex because they are unknown mixtures of DNA and other compounds. Thus, procedures for target DNA amplification from the environment require special attention.

The PCR has allowed us to go beyond the need for culturing prior to analyses of microbial communities. It has been shown that even microorganisms that can be routinely grown in the laboratory undergo some physiological changes when exposed to the environment. One of these changes (first observed by R. Colwell and colleagues) is known as the viable-but-nonculturable state and seems to be a common occurrence. Thus, the use of culture techniques paints only part of the picture in terms of microbial behavior under environmental conditions. The ability to amplify nucleic acids by the PCR has brought about a myriad of very ingenious modifications to the technique that can then be used to study complex ecosystems. The manner in which PCR can be modified is only limited by the need and/or imagination of the researcher.

This book presents several protocols that have been designed to amplify target sequences present in environmental samples. Since no single protocol will apply to all types of samples, it would be pretentious to assume that this manual covers all environmental applications of PCR. However, the manual will allow the reader to adapt protocols in order to fit research needs.

The chapters were written by active researchers who have published extensively on the use of PCR. Chapter 1 gives an extensive review of the current literature. Chapter 2 gives a general view of DNA and RNA amplification. In Chapter 3, Louws and colleagues share their experience with repetitive, sequence-based DNA amplification. Additionally, this chapter gives a very good sequence of methods for the isolation of DNA from different sources in order to obtain amplifiable material. Although it has been shown that thermostable DNA polymerases are rather hardy, they can be inhibited by several compounds, such as heavy metals. The presence of inhibitory substances has to be one of the most frustrating aspects of nucleic acid amplification in environmental samples; several chapters cover the elimination of such substances.

Bioaerosols are discussed in Chapter 7. This topic is receiving a lot of attention because of pathogenic microorganisms, such as *Mycobacterium* and *Legionella*. The amplification and detection of the latter microorganism is dealt with in Chapter 8. Readers may remember the recent *Legionella* outbreak on a cruise ship. The PCR amplification studies are giving a hint of the real importance of air as a vector for microorganisms. Legislation will have to be passed regarding air quality in the workplace. However, current guidelines are based on little information. It is very likely that PCR detection will allow for a more realistic approach to air quality.

Pepper and colleagues provide easy-to-follow protocols on the use of PCR for the detection of bacteria and enteric viruses in soils and sludges. Little is known about the incidence of these microorganisms in these environments. The use of sewage sludges to amend soils is a widespread practice, which, under some circumstances, may be a health hazard. Similarly, Chapters 5 and 6 deal with the detection of enteric viruses and the protozoan parasites *Giardia* and *Cryptosporidium* in waters. These pathogens have been implicated in several waterborne gastroenteritis diseases. Microorganisms capable of causing waterborne disease are present in waters at extremely low concentrations; thus, PCR offers a fast and sensitive solution: although it may not tell us much in terms of the infectivity status of the microorganisms in question, it nonetheless allows for quick monitoring of drinking waters.

Chapter 9 covers the amplification of ancient DNA, which is currently one of the most exciting topics in biology. The ability to isolate and amplify ancient DNA may not give us a dinosaur theme park, but it is giving us a glimpse of evolution and genes from extinct species.

There have been several excellent books and manuals written on the subject of nucleic acid amplification. A list of these is included in the section "Basic Equipment for a Nucleic Acids Amplification Laboratory." The reader is encouraged to check these other sources conjointly with this book in order to come up with a working PCR protocol. These will provide solutions and will limit unrealistic expectations of a technique, which, although powerful, is susceptible to extremely complex problems.

These chapters will provide a useful tool for the beginner as well as the seasoned researcher, and readers will benefit from the painful prior experience of the authors.

The editor wishes to acknowledge the help and encouragement of several colleagues (too many to mention by name, although C. Fricker and P. Bayman come to mind). This manual arose from several long nights of nonstop conversations at scientific meetings accompanied by copious amounts of a biotechnological product produced by a yeast, a type of which Raul Cano reportedly isolated from fossils. At future meetings, readers of this book may come up with new ideas and protocols while sipping metabolites of Jurassic-era yeasts.

PCR is one of the most powerful and useful techniques to be developed in the last decades. Nucleic acids can be amplified from any material if consideration is given to the complex nature of each sample. The authors exhort the reader to take this manual as a point of departure to modify protocols in order to get those precious, single-sharp bands on an agarose gel.

GARY A. TORANZOS

Contributors

NUMBERS in parentheses indicate the pages on which their contributions begin.

Morteza Abbaszadegan, Ph.D. (113, 129)
  Quality Control & Research Laboratory
  American Water Works Service Co., Inc.
  1115 South Illinois Street
  Belleville, Illinois 67220

Abdiel J. Alvarez, Ph.D. (37, 145)
  Abbott Diagnostics, Inc.
  P.O. Box 278
  Barceloneta, Puerto Rico 00617

Asim K. Bej, Ph.D. (1)
  Department of Biology
  University of Alabama at Birmingham
  251 Campbell Hall
  Birmingham, Alabama 35294-1170

Mark P. Buttner, M.S. (145)
  Harry Reid Center for Environmental Studies
  University of Nevada-Las Vegas
  Box 454009
  Las Vegas, Nevada 89154-4009

Frans J. de Bruijn, Ph.D. (63)
  MSU-DOE Plant Research Laboratory
  Michigan State University
  East Lansing, Michigan 48824

Raúl J. Cano, Ph.D. (183)
Biological Sciences Department
California Polytechnic State University
San Luis Obispo, California 93407

Ricardo DeLeon, Ph.D. (113)
Water Quality Division
Metropolitan Water District of Southern California
700 Moreno Avenue
La Verne, California 91750-3399

Elizabeth Fricker, Ph.D. (159)
Thames Water Utilities Ltd.
Spencer House Laboratory
Manor Farm Road
Reading, Berkshire RG2 0JN
United Kingdom

Charles P. Gerba, Ph.D. (95)
Department of Soil, Water and Environmental Science
429 Shantz Bldg., #38
College of Agriculture
The University of Arizona
Tucson, Arizona 85721

Frank J. Louws, Ph.D. (63)
MSU-DOE Plant Research Laboratory
Michigan State University
East Lansing, Michigan 48824

Ian L. Pepper, Ph.D. (95)
Department of Soil, Water and Environmental Science
429 Shantz Bldg., #38
College of Agriculture
The University of Arizona
Tucson, Arizona 85721

Teresa Picone, Ph.D. (159)
Roche Molecular Systems
1145 Atlantic Avenue
Alameda, California 94501

Maria Schneider, M.S. (63)
MSU-DOE Plant Research Laboratory
Michigan State University
East Lansing, Michigan 48824

Linda D. Stetzenbach, Ph.D. (145)
Harry Reid Center for Environmental Studies
University of Nevada-Las Vegas

Box 454009
Las Vegas, Nevada 89154-4009
Timothy M. Straub, Ph.D. (95)
Current Address:
Lookheed-Martin
Environmental Services Assistance Team, Region 10
7411 Beach Drive East
Port Orchard, Washington 98366
Gary A. Toranzos, Ph.D. (1, 37, 145)
Department of Biology
P.O. Box 23360
College of Natural Sciences
The University of Puerto Rico
Rio Piedras, Puerto Rico 00931-3360
Theresa Young (159)
Roche Molecular Systems
1145 Atlantic Avenue
Alameda, California 94501

Basic Equipment for a Nucleic Acids Amplification Laboratory

MANY laboratories around the world are applying molecular techniques to environmental problems. Many of these labs already have several years of experience and the necessary equipment to handle nucleic acids. A major consideration before embarking on a major purchasing expedition should be whether the laboratory will be actively involved in PCR and/or nucleic acid-based techniques or if the use of the techniques will be sporadic. The PCR is still a rather new technique, which, especially when dealing with environmental samples, can be very frustrating and expensive. It has been estimated, for example, that to carry out a typical smear test for the diagnosis of tuberculosis costs approximately $1, while a PCR diagnosis costs $11 [7]; thus, the more traditional techniques may still be desirable under some circumstances. If the technique is to be used routinely, then the researcher/laboratory manager should consider purchasing equipment of the highest quality possible, which needs a rather large original investment of money, time, and laboratory space. However, labs that will need these services infrequently should consider either contracting the work to a reliable laboratory or using some other laboratory that is fully equipped and has the necessary expertise in PCR.

Environmental laboratories should arrange to have physical space and some basic equipment, such as micropipettors dedicated strictly to PCR work. It is necessary to carry out all preparations in one area separated from the amplification area in order to avoid any chance of contamination by PCR products or primers, which can be devastating as a result of the exquisite sensitivity and efficiency of the technique.

This chapter lists the most basic equipment needed for PCR work. However, reasons for doing PCR amplifications vary; laboratory personnel should consider the aim of the laboratory and plan accordingly. Mark

Bloom and colleagues from the DNA Learning Center at Cold Spring Harbor Laboratories [1] have designed a very simple experiment to teach even high school students how PCR works. This is an amplification reaction based on only two temperatures, a beaker with boiling water, and a 55°C water bath. This does not "push the limits of the technique" but it does, nonetheless, allow for the amplification of a piece of target DNA. Thus, there exists the possibility of carrying out an amplification reaction with the minimum of equipment. However, although this may work well in a teaching laboratory, those may not be the best conditions for the reproducible and sensitive amplifications necessary when carrying out environmental work. The latter is especially true for work that has a bearing on public health decisions.

BASIC EQUIPMENT

(1) Thermal cycler: Choose this most basic equipment with extreme caution. Although experience has been good when using different cyclers, there is always the worry about access to technical services. Perkin-Elmer is the most widely used brand, and they manufacture a variety of other laboratory equipment and, therefore, have technical service departments in most countries. This is something to be considered by researchers who live in areas with limited access to these services. No specific brand is recommended, but under some circumstances the researcher might consider buying water-cooled cyclers over Peltier-type cyclers. The water-cooled cycler has some advantages in that it is mechanically simpler. However, these types of cyclers have limitations: they cannot go below the ambient temperature of the water without a water cooler, which increases the cost. Most of the cyclers are reliable. However, before buying one, the researcher should rely on the experiences of other colleagues. Some protocols are hard to standardize, thus, it is difficult to obtain consistent results if different brands are being used with the same protocol.

(2) Freezer: Choose a freezer ($-20°C$) without an auto-defrost or frost-free capacity. All enzymes and other materials used in the PCR are susceptible to changes in temperature. Frost-free freezers periodically heat up to melt the ice and should not be used. The frost helps keep the temperature low when opening and closing the door.

(3) Autoclave: All thermostable materials should be sterilized, and the autoclave should be periodically checked using an appropriate indicator, such as *Bacillus stearothermophilus* spores.

(4) Water still/deionizer: Only the most purified autoclaved water should be used to prepare the solutions.

(5) Incubators: Either water-jacketed or air-heated incubators can be used.

(6) Centrifuges (high speed and microfuges): These are useful for harvesting cells from pure cultures and/or for concentrating large water samples. Microfuges are used for separating the aqueous and organic phases during phenol/chloroform extractions of nucleic acids.

(7) Spectrophotometer and quartz (or UV-compatible plastic) cuvettes: These are used for quantitating primers and total DNA whenever necessary. The quantitation of the primers is of absolute importance in order to get consistent amplifications. Equimolar concentrations need to be used. Alternatively, each primer should be run in an agarose gel conjointly with a molecular weight marker, stained with ethidium bromide, and compared to each other to make sure that the concentrations are correct prior to use.

(8) Micropipettors (0–20 μl, 10–100 μl, 100–1000 μl, preferably auto-clavable): It is advisable (although not absolutely necessary) to use electronic ultramicropipettors, since these are less susceptible to human error when dealing with volumes less than 1 μl. Additionally, the use of plugged pipette tips decreases the probability of cross-contamination. Preferably two sets should be accessible (one set for pre-PCR and a second for post-PCR pipetting).

(9) Vortex mixers

(10) pH meters

(11) Magnetic stirrers

(12) Electrophoresis power supplies and gel boxes: A medium-voltage power supply (0–250 constant voltage) should be chosen. Additionally, horizontal gel boxes of different sizes should be accessible. Mini gels are needed for rapid amplicon visualization, while a midi or large gels are needed to separate complex amplicon patterns in cases such as RAPDs or REP-PCR.

(13) Laminar flow hood or still air enclosure: This prevents the spread of aerosolized target DNA.

(14) UV Transilluminator (254 nm or 366 nm wavelength): The shorter wavelength allows for the detection of smaller amounts of DNA in a band. However, this also damages the DNA if it is going to be used for other purposes. Some laboratories routinely use the transilluminator to expose microfuge tubes and pipette tips for five minutes prior to use in order to eliminate any contaminating DNA.

(15) Polaroid MP4 camera or equivalent (with an orange Kodak Wratten #23A filter): Polaroid-type 665 or 667 film or equivalent (ASA 3000).

PCR PATENT INFORMATION

The reader should be advised that PCR is a patented process (U.S. Patent #4,683,202; 4,683,195; 4,800,159; 4,965,188) and thus permission needs to be granted by Roche Molecular Systems, Inc., Atlantic Avenue, Alameda, CA 94501. Purchasers of Perkin-Elmer PCR reagents obtain a limited license for research and development purposes. Permission needs to be obtained under all other circumstances.

SUPPLEMENTARY LITERATURE

1 Bloom, M. V. 1994. Polymerase Chain Reaction. *Carolina Tips.* 57:14–18.

2 Berger, S. L. and A. R. Kimmel (eds.). 1987. *Guide to Molecular Cloning Techniques. Methods in Enzymology, Volume 152.* Academic Press, Orlando.

3 Eeles, R. A. and A. C. Stamps. 1993. *Polymerase Chain Reaction (PCR)—The Technique and Its Applications.* MBIU, R. G. Landes Co., Austin.

4 Erlich, H. A. (ed.) 1989. *PCR Technology, Principles and Applications for DNA Amplification.* Stockton Press, New York.

5 Innis, M. A., D. H. Gelfand, J. J. Sninsky and T. J. White (eds.) 1990. *PCR Protocols: A Guide to Methods and Applications.* Academic Press, San Diego.

6 Kricka, L. J. (ed.) 1992. *Nonisotopic DNA Probe Techniques.* Academic Press, San Diego.

7 Lewis, R. 1994. Polymerase chain reaction technology reigns from moose meat to microbes. *Gen. Eng. News.* 14:8–9.

8 Mullis, K. B., F. Ferré and R. A. Gibbs (eds.) 1994. *The Polymerase Chain Reaction.* Birkhäuser, Boston.

9 Towner, K. J. and A. Cockayne. 1993. *Molecular Methods for Microbial Identification and Typing.* Chapman & Hall, London.

10 Trevors, J. T. and J. D. Elsas (eds.) 1995. *Nucleic Acids in the Environment, Methods and Applications.* Springer-Verlag, Berlin.

11 Akkermans, A. D. L., J. D. van Elsas and F. J. de Bruijn (eds). 1995. *Molecular Microbial Ecology Manual.* The Netherlands: Kluwer Academic Publishers.

CHAPTER 1

Introduction

ASIM K. BEJ[1]
GARY A. TORANZOS[2]

SINCE the advent of molecular methods for the manipulation of nucleic acids, great progress has been made in the understanding of the genetic makeup, taxonomy, species diversity, distribution, occurrence, and community structure of microorganisms in the changing environment. The milestones in the discovery of various molecular methodologies have advanced the knowledge of cellular processes and continue to revolutionize the understanding about the importance of microorganisms in the environment. As a result, molecular techniques have become the key component in the study of gene regulation in environmentally important microorganisms. Monitoring microorganisms (including pathogens) and their activities in the surrounding environment has always been an important area of investigation. The rapidity, dependability, and the quality of information that can be obtained by following genetic technologies have made many of the molecular approaches the "gold standards" in microbiology.

During recent years, it has been realized that a conventional culture-based survey of microorganisms in environmental samples, using their physiological characteristics, has several disadvantages. Culturing in nonselective media followed by confirmation tests in selective media is time-consuming. For microbial community analysis and microbial interactions in the environment, the nonselective medium used for their recovery may actually be the selective medium for many of the species. As a result, when an environmental sample is analyzed by traditional culture tech-

[1]Department of Biology, University of Alabama at Birmingham, Birmingham, AL 35294-1170, U.S.A.
[2]Department of Biology, P.O. Box 23360, The University of Puerto Rico, Rio Piedras, Puerto Rico 00931-3360.

niques, only a fraction of the total microbiodiversity is recovered; therefore, the actual microbiodiversity is greatly underestimated.

As stated by Gutell, et al. [1a], the knowledge of the real microbial diversity was extended to a catalog of anecdotal facts. It is a recognized fact that there are many more microorganisms than can be isolated, thus, studying them becomes extremely difficult. Studying microbiodiversity gives a clue about the role of microbiota in the biogeochemical cycles and strategies of survival of microbial groups, and possibly the most obvious is that the microorganisms have unique enzymes that can be used for the production of desirable products.

Some benefits derived from studies on microbial activities and their roles in a given ecological community are the use of microbial activities, such as biodegradation of xenobiotics, protection of crops from pests, symbiotic relationships that help to form a healthy ecological balance among various life forms, and the protection of human health from microbial pathogens. Also, it is known that the cellular physiology with the underlying genetic regulations is not quite the same in natural environments as when cultured on artificial microbiological media.

The lack of appropriate methodologies to study microbial activities in situ may be somewhat overcome by the advancement of DNA-based techniques. Part of this success is due to the combination of one or more molecular methodologies to increase the reliability of the overall approach. The DNA-based, noncultural detection of microorganisms in various environmental samples has overcome problems associated with the traditional microbiological culture methods. Although these DNA-based techniques to study the microbial activities at a molecular level have been utilized in pure cultures and specifically in medically important microorganisms, their applications to environmentally related microorganisms have just recently been used. This is in part due to the complex nature of the environmental samples from which the recovery of sufficiently pure nucleic acids for conducting various molecular processes was a challenge to the investigators.

However, the technological advancements have made it possible to recover the nucleic acids from complex environmental matrices for applying the DNA-based methodologies. Isolation and identification of the target microbial population, which comprises a minor proportion of the total microbial community, from the complex environmental matrix requires either culturing of the microorganisms by conventional microbiological methods or recovering nucleic acids from the environmental sample followed by removal of interfering factors so that molecular techniques can be successfully applied. The development and subsequent applications of techniques for the extraction and purification of DNA and RNA from the complex environmental matrices and detection of a target microbial population by identifying their nucleic acids by gene probes or polymerase chain reaction (PCR) methods have provided us with information in the areas of microbial ecology, public health indicators of microbial contamination, and bioremediation.

Furthermore, genetic manipulations in many of the naturally occurring microorganisms, their persistence, fate, and ability to perform the desired tasks in the environment, such as enhanced food and agricultural production, metal and mineral leaching, biopesticide/biocontrol, and waste treatment, have already shown great potential and undoubtedly will revolutionize the environmental microbial sciences in the near future.

The invention of a technique for the amplification of nucleic acids [1] was nothing short of revolutionary. The development of the PCR amplification idea, as told in the already well-known anecdote of Kari Mullis and his sleepy girlfriend on a highway in Mendocino County, set molecular biology aboard a scientific equivalent of the *Train à Grande Vitesse*. The technique has become an intrinsic part of the routine application in many laboratories.

The PCR is an *in vitro* method for replicating defined DNA sequences of specific organisms by using their DNA as templates. The amount of target sequence is increased exponentially by using PCR. While previously only minute amounts of a specific gene could be obtained from a cell, now even a single gene copy can theoretically be amplified to 1 million copies within a few hours. The PCR consists of repetitive cycles of DNA denaturation to convert double-stranded DNA to single-stranded DNA, annealing of oligonucleotide primers to the target DNA sequences, and extension of the DNA by incorporation of the corresponding deoxynucleotides by the action of a DNA polymerase [2], resulting in an exponential increase in the amount of the target DNA.

Nucleic acid amplification by the PCR is a very powerful technique when used appropriately. It would have been unheard of a few years ago that it would actually be possible to isolate genes from ancient (and even fossilized) materials. However, as it often occurs, many people have lost track of the border between science fact and science fiction. The book *Jurassic Park* (and the movie by the same name) did wonders to introduce PCR to the lay public. This seems to have awakened unusual expectations that have even touched the scientific community. Many of us have been approached with the typical question regarding the possibility of reviving dinosaurs or other extinct life forms.

Recently, in a real experiment, a microbial species, *Aspergillus*, was detected from the lung tissues of a 5,300-year-old iceman using PCR amplification of 18S rRNA as a target [3]. In another study, in a portion of the 16S rDNA from an ancient *Bacillus* spp., which was trapped in bees 25–40 million years ago in Dominican amber, the PCR was amplified, cloned, and sequenced [4].

It is important to be aware that the PCR is in fact a very powerful technique. However, it is equally important to know its limitations. The application of this technique in the identification and characterization of ancient DNA has been elaborated in Chapter 9.

Well over 20,000 research articles have been published in which PCR was used in one way or another, and the number of publications are increasing every year. Practically every branch of the biological sciences is using this amplification technique. Needless to say, environmental microbiology has also been touched by this "gold rush." The application of the PCR DNA amplification method to environmental sciences has shown great potential for solving many of the problems that previously could not be overcome by conventional microbiological culture methods or by traditional gene-probe or nucleic acid hybridization methods.

The initial report on the use of amplification techniques for environmental microbiology was presented *ad libitum* by R. M. Atlas and R. J. Steffan at the REGEM I (Release of Genetically Engineered Microorganisms, 1988) meeting in Wales, Great Britain. They described PCR use for tracking a genetically engineered *Pseudomonas cepacia* in soil samples. Since that short report, PCR has been used in many different areas of environmental science. A good example of this is the large number of papers presented at the Seventeenth Biennial Conference of the International Association on Water Pollution Research and Control (Budapest, Hungary, July 24–30, 1994) where at least twelve of the presented papers in the Health-Related Water Microbiology Section used PCR amplification for the detection of pathogens in waters. Several others at least mentioned the use of PCR.

The PCR DNA amplification method has many advantages over conventional methods for its application in the environmental sciences [5,6]. Some of the recent modifications to PCR methodology have tremendously helped the application of PCR to environmental samples. Some of these are: the "hot start" that increases the specificity of the reaction, the randomly amplified polymorphic DNA (RAPD) that allows for fingerprinting DNA, and *in situ* PCR. Various other developments have included the isolation of thermostable DNA polymerases from various thermotolerant microorganisms, some of them with additional activities, such as reverse transcriptase (Tth DNA polymerase from *Thermus thermophilus*), which is specifically important to study *in situ* gene expression in the environment [5–9], that may be very useful to study the microorganisms in the environment. General considerations and methodologies for PCR amplification are described in later chapters.

It should be realized that, although the amplification technique seems simple, it has some intrinsic problems and idiosyncracies. The first and the most important aspect is that, when working in a molecular biology laboratory, most of the samples being analyzed consist of highly purified

DNA or pure cultures of organisms. However, it has been reported that even toothpicks used for colony transfers can inhibit PCR amplification [10]. In most environmental samples the target DNA is present at extremely low (and dirty) concentrations. This is obviously the case where PCR is and should be the method of choice. However, all samples need to go through a purification process prior to PCR amplification. A general approach for the identification of microorganisms and for studying their activities in the environment using PCR-based *in vitro* DNA amplification methodologies is schematically represented in Figures 1a and 1b.

RECOVERY AND PURIFICATION OF NUCLEIC ACIDS FROM ENVIRONMENTAL SAMPLES

The first step in the application of PCR methodologies in environmental samples is the recovery of DNA. Removal of various inhibitors, such as humic materials, proteins, and heavy metals, is essential for successful application of PCR DNA amplification. Primarily, two approaches have been reported to recover DNA from environmental samples: (1) isolation of microbial cells followed by lysis and purification of the nucleic acids

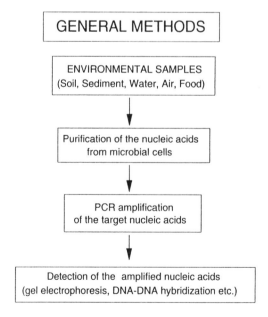

Figure 1a Schematic representation of a generalized approach for in vitro nucleic acid amplification to study microorganisms in the environment.

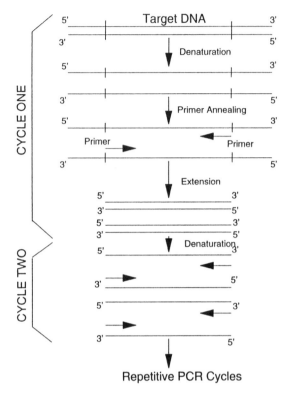

Figure 1b Schematic representation of the conventional PCR amplification approach. Target DNA is denatured at an elevated temperature followed by primer annealing at a relatively lower temperature to its target DNA. The temperature is elevated to an optimum level to extend the primer by synthesis of the strand using a thermostable DNA polymerase. This cycle is repeated at least twenty-five times to achieve 1 millionfold amplification of the target DNA or genes.

(cell-extraction method) [11], and (2) direct lysis of the microbial cells in the environmental matrix followed by nucleic acid purification (direct-lysis method) [12]. Detailed procedures for the removal of PCR-inhibitory substances from various complex environmental samples are described in Chapters 4, 5, and 6.

PURIFICATION OF DNA FROM SOIL AND SEDIMENTS

Purification of the released DNA either by direct-lysis or cell-extraction methods is performed by applying a combination of the various standard purification methods, such as phenol-chloroform extraction, treatment with ammonium acetate followed by ethanol precipitation repeated polyvinylpolypyrolidone (PVPP) treatment or dialysis, hydroxylapatite or affinity chromatography, and multiple cesium chloride density gradient centrifugation [7,13]. In a separate approach, treatment with lysozyme, followed by repeated freezing and thawing of the samples to release nucleic acids from soil samples, has been described by Tsai et al. [14]. The released nucleic acids were then purified by standard phenol-chloroform extraction and chromatography methods. Using this approach, it was possible to detect < 10 cells of *Escherichia coli* per 1 g of seeded or unseeded soil samples [14–16] (see Chapter 4 for a complete protocol).

In a similar study, bacterial cells were differentially separated from soil colloids on the basis of their buoyant densities [17]. In this method, a modified sucrose gradient centrifugation protocol was developed to separate most of the soil colloids from the bacterial cells in the sample for PCR amplification of the target DNA. However, since this approach retained minute quantities of inhibitory colloidal soil particles in the bacterial cell suspension, it was necessary to run an additional twenty-five cycles of PCR DNA amplification, with an aliquot of the amplification product of the first twenty-five cycles ("double PCR"), to detect one to ten target microorganisms per gram of soil sample. In a separate study, DNA was extracted and purified from soil using cold lysozyme and SDS-assisted lysis with either freeze-thawing or bead beating, cold phenol extraction of the resulting soil suspension, CsCl and potassium acetate precipitation, and spermine-HCl or glass milk purification [18]. The resulting DNA was pure enough for PCR amplification of up to 20 ng of soil-derived DNA from *Pseudomonas fluorescens* (RP4::*pat*) per 50 μl reaction mix.

Recently, an effective method for removal of inhibitory humic substances in crude DNA extract from soil samples was described by using Sephadex G-200™ spun columns (Pharmacia LKB Biotechnology, Inc.) [14–16]. This extraction method resulted in the detection of < 70 cells of *E. coli* by PCR DNA amplification method from a fraction of the purified sample using 16S rRNA as the target.

Also, a simple and effective method of purification of PCR-amplifiable DNA from soil matrices has been described by separating crude DNA extracts by direct-lysis method in a low-melting point agarose gel mixed with polyvinylpyrrolidone (PVP) [17]. The water-soluble PVP forms hydrogen bonds with phenolic compounds in the DNA extracts and prevents comigration during agarose gel electrophoresis. Differentially migrated DNA bands are purified from the agarose gel.

A more tedious and long procedure for extraction of PCR-amplifiable DNA from soil, sediment, and sand samples for the detection of native bacterial populations using 16S rRNA genes and mercury resistance gene (*mer*) has been described [18a,19]. This procedure required lysis of the cells by treatment with sodium dodecyl sulfate followed by concentration of the nucleic acids with polyethylene glycol and NaCl. After CsCl density gradient centrifugation purification, the DNA was further purified by phenol:chloroform treatment. The DNA was then used for PCR amplification. The entire procedure for the extraction and purification of DNA takes three days to complete.

A slightly different approach has been described for purification of total DNA from humic acid-containing soil samples and the effect of the presence of humic acids in the purified DNA on molecular approaches, such as PCR and hybridization, has been evaluated [20]. In this approach, in addition to lysozyme, another cell lysing enzyme, lyticase (Boehringer Mannheim) was used. Following lysis, the sample was purified by phenol-chloroform-isoamyl alcohol. Additional brownish-colored humic acids associated with the DNA sample were purified by passing the sample through an ion exchange column (Qiagen-Tip 500). The purified DNA could be used for DNA-mediated genetic transformation of *E. coli*, slot-blot hybridization, or PCR DNA amplification.

In another approach, total DNA was purified from low quantities (100 mg) of soil samples seeded with bacterial species, such as *Agrobacterium tumifaciens* and *Frankia* spp. In this extraction procedure, 100 mg of soil sample was resuspended in a buffer and sonicated. The sample was then "freeze-boiled" (liquid nitrogen and boiling water) to lyse the cells and the DNA purified in an Elutip™ column (Schleicher & Schuell) for final PCR amplification. The detection of the target microorganisms had a sensitivity of $10^3 - 10^7$ cells of *A. tumifaciens* and 0.2×10^5 cells of indigenous *Frankia* spp. per gram of soil sample [21]. Purification of total DNA from humic and other inhibitory colored substances has been described by a simple electroelution approach. This approach takes only about fifteen minutes and 50–60% of the DNA from the sediment can be recovered and used for PCR amplification [22].

PURIFICATION OF NUCLEIC ACIDS FROM MICROORGANISMS IN WATER

The isolation and purification of nucleic acids from microorganisms present in water, in general, can be trouble-free, and primarily depends on the type of water. A simple and rapid method for isolating nucleic acids from aquatic samples was described by Sommersville et al. [23]. In this approach, 300 to > 1000 ml of water sample was concentrated on a single cylindrical filter membrane to harvest the cells, followed by alkaline lysis and proteolysis performed within the filter housing. Crude, high-molecular-weight nucleic acids were extruded from the filter unit and further purification was carried out by conventional methods, such as CsCl density gradient centrifugation or by treatment with NH_4-acetate. Using this approach, it was possible to extract sufficiently pure chromosomal

DNA, plasmid DNA, or various species of RNAs (5S, 16S, and 23S). This rapid and convenient approach, using filtration to concentrate relatively large volumes of environmental waters for recovery of dissolved and particulate DNA, can be used for PCR-based analysis of community diversity and microbial activity in aquatic ecosystems. In various other studies, CsCl-EtBr density gradient centrifugation [24] (for environmental DNA), multiple PVPP treatment (for rRNA) [25] repeated phenol-chloroform extractions (for total planktonic DNA or cyanobacterial DNA) [26–28] were found to successfully yield sufficiently pure DNA.

In a similar approach for the detection of microbial cells (both indicators and pathogens) in drinking water or relatively clear waters, cells are concentrated on a PCR-compatible membrane filter (e.g., Teflon), lysed by repeated freeze-thaw method followed by PCR DNA amplification without removing the filter from the reaction tube [29,30]. Alternatively, filtered cells can be lysed directly on the polycarbonate filter by lysozyme, and the released DNA can be purified by phenol-chloroform extraction followed by ethanol precipitation to yield adequately purified DNA for PCR amplification and analysis [31,32]. Optimization of spin time and speed have been evaluated for Sephadex G-200 columns for purification of bacterial DNA from water samples [33].

Although the isolation and purification of DNA from soil and sediment is summarized in Figure 2, and elaborated in Chapter 4, a standard protocol for removing all possible inhibitors, including the humic acids in various types of soil and sediment samples that can be utilized for PCR amplification and gene probe analysis, is still an open area of investigation. It should be noted that environmental compounds may poison the enzyme or may lower its V_{max}. Yet other compounds may simply bind to the DNA making it inaccessible to the polymerase.

PURIFICATION OF RNA FROM ENVIRONMENTAL SAMPLES

Application of PCR technology has a special value in detecting RNA from target microorganisms in the environment. Information from genetic and physiological research suggests that many bacteria possess a wide variety of genes that are not expressed in the environment [34]. Moreover, many of the genes, which are expressed in the laboratory, may not be well expressed in the environment due to stress conditions. Therefore, to study the expression of a gene of interest, specific mRNA rather than DNA should be used as the target. Extraction of rRNA/tRNA, followed by northern hybridization to determine the species composition in a marine sample, has been evaluated [35]. A generalized protocol for the extraction and purification of RNA from environmental samples is summarized in Figure 3. Although it is a relatively new approach, the use of gene probes

HOMOGENIZE ENVIRONMENTAL SOIL OR SEDIMENT
Transfer 100 mg soil or sediment into a 1.5 ml microcentrifuge tube
Add 350 μl homogenization solution A [250 mM NaCl+100 mM Na₂EDTA, pH 8.0]
Vortex/grind for 30 sec and sonicate for 3 min

CELL LYSIS
Add 100 μg lysis solution B [250 mM NaCl+100 mM Na₂EDTA, pH 8.0+4% SDS]
+
NovoZym® (Novo Biolabs, St. Louis, MO)

CONCENTRATE DNA
Add 750 μl isopropanol to supernatant liquid
30 min at -20°C
Centrifuge 12,000 x g for 15 min
Suspend dried pellet in 100 μl water

PURIFY DNA
Separate by electrophoresis in 1% low melt point agarose
Melt agarose block and add 3 volume water and incubate at 68°C for 20 min

USE DNA DIRECTLY FOR PCR AMPLIFICATION

Figure 2 Schematic representation of general methods for recovery, lysis, and PCR amplification of a target microbial pathogen, which is followed by colorimetric oligonucleotide probe hybridization detection of the amplified DNA on solid supports.

Collect cells on high capacity cartridge-type filters with the diameter
of 142 mm and freeze in dry ice (minimum 10^9 cells may be required)
(0.22 μm pore size Durapore or Sterivex -GS cartridge, Millipore)
↓
Five ml of SDS in STE buffer which is preheated to 85°C was added
to the filter and DEPC treated water added to a final concentration
of 0.1%
↓
The sample is boiled for 5 min with intermittent vortexing
↓
The liquid is transferred to a new tube and kept in ice
↓
Additional 5 ml of STE-SDS in DEPC treated water is added to the
filter, vortexed and the liquid is combined to the previous sample
↓
Ten ml of GIPS, 1 ml of 2 M NaOAc (pH 4.0) + 10 ml phenol and 2 ml
$CHCl_3$ (49:1) is added to the samples and mixed well
↓
Following centrifugation at 4°C, the supernatant is treated with
glycogen and precipitate 2 x with isopropanol
↓
The RNA pellet is washed with cold 70% alcohol
↓
The pellet is dried and resuspended in 1 mM EDTA
↓
The sample is treated with 0.1X volume of 2 M NaCl and 0.7X volume
of isopropanol
↓
The pellet is washed 1X with cold 70% alcohol, dried and
resuspended in DEPC-treated water or Tris·EDTA (pH 8.0) buffer

Figure 3 An outline of the generalized protocol for the extraction and purification of mRNA from environmental biomass using the boiling-lysis method.

against a target mRNA in the environment has the potential to provide us with information about the gene activity, physiological state, and possibly, viability of microorganisms in the environment.

An additional benefit of targeting mRNA instead of DNA is that it is usually found in high copy number. It should be noted that the half-life of the majority of the bacterial mRNA is very short (a few minutes). Therefore, recovery of complete transcripts is difficult. Presence of multiple copies of the transcripts and blocking the RNAse synthesis by pretreating the cells with chloramphenicol helps to recover intact transcripts. Moreover, the high sensitivity of the PCR approach may not require high copy number of the target RNA/mRNA from the target microorganisms.

For the isolation and purification of total RNA from microorganisms in the environmental samples, a soil sample seeded with *Pseudomonas aeruginosa* PU21 was treated with guanidium thiocyanate mixed with sodium citrate, sarcosyl, and 2-mercaptoethanol to achieve lysis of cells, fixation of total cellular RNA, and hydrolysis of DNA. The treated sample was then purified with phenol-chloroform-isoamyl alcohol yielding 17 μg of total RNA and 0.16 μg of mRNA from 1 gram of soil containing 8×10^8 *P. aeruginosa* PU21 cells. This extraction method can be completed within a few hours and has the potential to allow the study of microbial gene expression in the environment by reverse transcription of the target mRNA followed by PCR DNA amplification with specific primers sets. More recently, lysozyme-hot phenol treatment followed by gel filtration with Sephadex G-75 spun columns was used to recover total rRNA from microbial communities in sediment, soil, and water samples [36]. The purified total RNA from various environmental samples has been shown to be useful for molecular procedures, such as hybridization.

Purification of mRNA from water samples that is encoded by the *mer*A gene in *P. aeruginosa* upon induction has been reported [37]. In this approach, the cells were collected on a filter and lysed by freeze-thaw.

PURIFICATION OF VIRAL NUCLEIC ACIDS FROM ENVIRONMENTAL SAMPLES

Reports of outbreaks of enteric viral diseases due to drinking contaminated water have increased significantly during the past decade [38,39]. Conventional monitoring for the presence of enteric viruses in the environment and drinking water requires animal cell culture, is technically difficult, and time-consuming. Since enteric viruses are relatively resistant to water treatment and disinfection processes, a rapid and efficient method for periodic, perhaps routine monitoring of the drinking water and finished water sources for possible viral contamination, is necessary. Detection of enteric viruses from water samples requires sample cleanup for successful PCR DNA amplification.

If water samples are to be analyzed for the presence of a specific pathogen, such as an enteric virus, the water is usually processed by a process known as Viradel (Virus-Adsorption-Elution), which uses a filter matrix to concentrate viruses electrostatically. This process not only concentrates viral particles, but also any particle present in the water that may have similar characteristics to the viral particles. In addition, large clumps can be trapped by the filter, thus, many unwanted compounds end up in the final concentrate. In this process, the very same compounds used for the elution of the viral particles from the filter matrix [39] may interfere with the amplification of the target nucleic acids.

Thus, the first step is to concentrate the viral particles followed by the processing of the samples to eliminate environmental contaminants that may potentially inhibit the PCR amplification reaction. After precipitation, the viruses from the filter can be eluted and RNA can be freed by phenol-chloroform treatment. The aqueous solution can then be purified using a Sephadex G-15, G-25, G-50, G-100, or G-200 spin chromatography column and further concentrated by a Centricon™ 100 (Amicon, MA) or a Microsep™ (Filtron, MA) microconcentrator. The retenate is then used directly for reverse transcriptase-PCR (RT-PCR) amplification of the target viruses [39]. By following this approach, 97% of the seeded enteroviruses were recovered from waters [39].

The presence of infectious RNA viruses in wastewaters, sludges, and soils, and their possible role in the spread of disease is a major concern. Total nucleic acids, directly from environmental samples or from viral concentrates from various sources, can be extracted and analyzed using the RT-PCR DNA amplification method. Recently, Ansari et al. [40] reported the detection of HIV-1 in domestic sewage and wastewater samples. However, the study did not show any evidence of the infectivity of these viruses in the wastewaters or their possible role in spreading disease to the human population. It is unlikely that there is risk associated with the wastewater for spreading HIV in the human population. However, these studies indicate that the prevalence of certain diseases in a population can be monitored by analyzing domestic sewage samples.

More recently, a procedure for purification of enteroviruses from groundwater samples and samples containing humic acids has been described [41]. In this approach the samples were treated with Sephadex 100 or 200 spin columns™ (Pharmacia) in combination with Chelex 100™ (BioRad Laboratories) to remove interfering factors for RT-PCR detection. This approach, for detection of enteroviruses by RT-PCR, effectively removed the contaminants, including the humic materials, from the water sample.

Tsai et al. [42] have described a simple ultrafiltration method to concentrate enteroviruses and hepatitis A viruses (HAV) from sewage and seawater samples. Using this purification approach, a triplex RT-PCR approach has been developed for simultaneous detection of poliovirus, hepatitis A virus, and rotavirus in sewage and ocean-water samples [42]. Using Sephadex G-50 and Chelex® 100 resin columns, the sewage sludge samples were purified for recovery of the viral RNA that was devoid of PCR-inhibitory substances. By following this purification approach, enteroviruses were detected in ten different sewage sludge-amended soil samples. Applications of PCR methodologies for recovery, purification of PCR-compatible target nucleic acids, and detection with high sensitivity and specificity by PCR methodologies have been elaborated in Chapters 4 and 5.

A different approach, called "antigen-capture PCR" (AC-PCR), has been described for the detection of hepatitis A (HAV) virus in environmental samples [43]. In this approach, the HAV was captured from seeded liquid wastes by homologous antibody, heat-denatured, and RT-PCR performed in a single reaction tube. The AC-PCR amplified products were detected by gene probe hybridization.

DETECTION OF WATERBORNE BACTERIAL PATHOGENS AND INDICATORS

Apart from the detection and monitoring of indicator microorganisms for water-quality assessment, it is also important to detect various water-borne microbial pathogens with high sensitivity and specificity. *Legionella* spp. is a water-related microbial pathogen and can be transmitted to humans via aerosols. The resulting cause is the condition known as Legionnaires' disease. Starnbach et al. [44] reported the detection of *Legionella pneumophila* by amplification of a fragment of DNA of unknown function. Mahbubani et al. [45] have developed a method, based on PCR and gene probes, for the detection of *Legionella* in environmental water sources. All species of *Legionella*, including all fifteen serogroups of *L. pneumophila*, were detected by PCR amplification of a 104 bp DNA sequence that codes for a region of 5S rRNA. A radiolabeled oligonucleotide probe complementary to an internal region of the amplified DNA was then used to confirm amplification. Strains of *L. pneumophila* (all serogroups) were specifically detected based on amplification of a portion of the coding region of the macrophage infectivity potentiator (*mip*) gene. A multiplex PCR-based colorimetric detection of the genus *Legionella* and *L. pneumophila* is described in Chapter 8.

A PCR gene probe detection of *Salmonella* spp. has been developed using the *hns* [46], *himA* [47], and *Salmonella* Plasmid Virulence (*spv*) gene (Bej, unpublished data).

Giardia lamblia causes waterborne gastroenteritis in humans. Using PCR amplification of different segments of the giardin gene of *G. lamblia*, it was possible to differentiate *G. lamblia* from *G. muris* [48]. Also, a single *Giardia* cyst was detected by PCR amplification after separating the cyst by a micromanipulator [49]. The use of immuno-capture PCR to selectively capture *G. lamblia* cysts in 100 gallons of concentrated river water by immuno-magnetic beads followed by PCR amplification showed high sensitivity (Mahbubani, unpublished data).

In *E. coli* β-D-glucuronidase is produced by the *uid*A gene. Bej et al. [50,51] developed a method for the detection of *E. coli* and *Shigella* spp. using four different regions of the *uid*A gene and part of the *uid*R gene, which is the regulatory region of the *uid*A gene, as targets. Besides being less time-consuming and having higher specificity and sensitivity, the most important advantage of this method over conventional methods is that *E. coli* that do not express the *uid*A gene can still be detected by the PCR, since the gene, although not expressed, may still be present in the bacteria.

Similarly, Cleuziat and Baudouy-Robert [52a] used a large region of the *uid*A gene of *E. coli* as a target for PCR amplification and gene probe detection of *E. coli* and *Shigella* spp. This PCR gene, probe-based method might have the specificity and sensitivity required for monitoring indicator organisms in environmental and potable waters. A field evaluation of the PCR application detection of enteric pathogens and indicator microorganisms has been reported using *uid*A and *lac*Z as targets [51]. Recently, the coding sequence of the *uid*A gene from *S. sonnei, S. flexneri, S. dysenteriae* and *S. boydii* has been sequenced and compared with the *uid*A gene of *E. coli*. Oligonucleotide primers for the *E. coli uid*A gene have been designed. These oligonucleotide primers could be used for the detection of indicator microorganism, *E. coli* for water-quality monitoring, as required by the U.S. Environmental Protection Agency. The application of gene probes and PCR amplification for detection of group-specific or species-specific microorganisms in water is rapid and shows great promise as a routine monitoring technique for microbial water quality.

SOLID-PHASE PCR AND ITS USE IN PATHOGEN DETECTION

Most PCR amplifications are carried out in solution. Generally, the volume of the PCR reaction is very small (100–125 μl) and requires concentration of the viruses or bacteria from large volumes of water prior to amplification. Conventionally, microorganisms are concentrated from environmental waters by membrane filtration (bacteria and parasites) or adsorption-elution (viruses). The efficiency of concentrating microorganisms from environmental samples is poor and ranges from 3–20% for parasites and from 50–80% for viruses. In addition, if the concentrated microorganisms are to be lysed to release the nucleic acids, followed by precipitation, at every step there is loss of target nucleic acids affecting the sensitivity of detection. It would be useful to concentrate the microorganisms on a membrane filter, release the nucleic acids and immobilize them on the membrane filter surface, and carry out the amplification procedures directly on the same solid surface.

Following this principle of cell collection on a membrane filter followed by lysis and covalent binding of the released nucleic acids on the membrane, PCR amplification detection of target microbial pathogens has been reported [52,53]. This "solid-phase" amplification lends itself for the simultaneous detection of several microbial pathogens present in water or air. In addition, since the target nucleic acids are covalently bound on the membrane filters, the same filters can be subjected to several rounds of amplification using different amplification conditions. Also, once the nucleic acids are immobilized, the membrane can be stored at room temperature for long periods of time until used again.

SIMULTANEOUS AMPLIFICATION OF MULTIPLE TARGETS (MULTIPLEX PCR)

It is possible that the environmental samples and drinking waters may contain more than one type of microbial pathogen in addition to the indicator microorganism. Use of multiplex PCR for amplification and detection of more than one target in a single PCR reaction can be useful for monitoring multiple microbial pathogens in a single environmental or water sample. This method was first described by Chamberlain et al. [54] to detect human genes. Modifications of this approach of simultaneous PCR amplification of multiple targets associated in different bacteria in the environmental samples has been demonstrated [29,32,50,55,56]. Multiplex amplification of two different *Legionella* genes, one specific for *Legionella pneumophila (mip)* and the other for the genus *Legionella* (5S rRNA), was achieved by staggered addition of two different sets of primers at two different concentrations [32]. By this method it is possible to detect members of the genus *Legionella* as well as *L. pneumophila* should the two different species be present in one sample.

In a water-quality monitoring study, simultaneous PCR amplification was performed using *lacZ* and *uidA* as targets. In this study, it was possible to detect in one sample total coliform bacteria by amplification of the *lacZ* gene, the indicator microorganism *E. coli*, and *Shigella* spp. by the amplification of the *uidA* gene [51]. A triplex, PCR assay using heat labile toxin (LT), shigalike toxin I (SLT I), and shigalike toxin II (SLT II) genes as targets was used for identification of toxigenic strains of *E. coli* in water samples [55a].

A multiplex PCR-based detection of another microbial pathogen, *Vibrio parahemolyticus*, which may be hemolytic [Kanagawa positive (K +)] or nonhemolytic [Kanagawa negative (K −)], has been used in environmental samples by using two different target genes, thermostable direct hemolysin (*tdh*) for K + and thermolabile (*tl*) for K − strains [56a].

Recently, a useful approach called DIAPOPS (detection of immobilized, amplified product in a one-phase system) for the detection of *Salmonella* and Bovine Leukemia virus has been reported using a microtiter plate, in which one of the two oligonucleotide primers was immobilized prior to the PCR amplification. Following amplification, the detection of the amplified DNA was detected colorimetrically in the same well in which the PCR amplification was performed [57].

In future studies, it may be desirable to group certain microbial pathogens and indicators in the environmental samples and design the primers for specific targets. For example, one can group all the waterborne and airborne pathogens and amplify all the specific target genes in a single reaction for their detection. When several genetically engineered microorganisms (GEMs) are released together for the degradation of complex hazardous wastes and pollutants, they can be monitored together in a single PCR reaction by amplifying a unique segment of the DNA from each of the GEMs, or together by amplifying a common segment of all the GEMs that is not present in other bacterial species.

There are several reports on the existence of many microorganisms, including human pathogens, in the environment in a viable, but nonculturable (VBNC), that is, dormant stage [49,56,58–62]. These microbial pathogens are shown to be potentially infectious when suitable conditions prevail [59]. One obvious difficulty in elucidating this potential hazard is the inability to detect these viable, but nonculturable, cells in the environment because the routine microbiological methods will not allow them to grow (on agar media) or will not distinguish them from the dead cells (by microscopic techniques). Recognizing that the terms *alive* and *viable* are subject to different definitions, the reasonably acceptable definition would be that the live cells are considered those capable of cell division, metabolism (respiration), or gene transcription (mRNA production) [60].

For the detection of those microbial cells that are in a viable, but nonculturable, state in the environment, it is desirable to target mRNA for cDNA synthesis followed by PCR amplification. The potential problem of this approach would be that most of the prokaryotic mRNAs have half-lives of only a few minutes. Mahbubani et al. [61] have shown that the mRNA of the *mip* gene of *L. pneumophila* can be stabilized simply by growing the cells for ten to fifteen minutes in the presence of chloramphenicol before harvesting. They have shown that the PCR amplification of the *mip* mRNA could be a potential means for the detection of metabolically active *L. pneumophila* cells. Use of chloramphenicol for increasing stability of bacterial mRNA is yet to be tested in other microorganisms.

Another perplexing issue that may create additional problems in such an approach is the efficiency of gene expression of these dormant microbial pathogens. It is possible that the transcription or regulatory systems of the target genes in these microbial pathogens are inhibited by various environmental factors and inhibitors when they are present in the natural environment. Therefore, in this situation the quantity of the target mRNA level may be so low that it may remain undetected even by PCR. Besides *L. pneumophila*, another important microbial pathogen, *Vibrio vulnificus*, that inhabits marine and estuarine waters can cause fatal infections when humans ingest contaminated raw oysters, has been found to enter in a viable, but nonculturable, state during the colder months and resuscitate from the nonculturable state when a suitable environment prevails during the summer months [58]. Using PCR amplification of the hemolysin gene, Brauns et al. [58] detected DNA from culturable and from nonculturable *V. vulnificus* cells. Although there was an apparent decrease in the sensitivity of detection of nonculturable cells by the PCR, it is not well understood at this time, and several possible explanations have been described [58].

Among these possibilities, the important criteria that may be of concern in applying the PCR methodology for the detection of viable but nonculturable microorganisms are (1) less DNA content per cell, (2) difficulty in breaking open because of changes in the cell wall that may occur due to carbon or nitrogen starvation or changes in the environmental conditions. However, Brauns et al. [58] did not attempt to use the hemolysin mRNA as a target for PCR amplification from the nonculturable cells, which could have determined the exact nature of the cells, that is, alive or dead.

A study by Mahbubani et al. [61] has shown that, since *Giardia* cysts killed by heat-treatment or monochloramination also give positive mRNA PCR amplification, mRNA-PCR alone is not sufficient to distinguish live from dead *Giardia* cysts. Therefore, in this organism, using the giardin mRNA as a target for PCR amplification, it is necessary to include an mRNA induction step to determine the viability of the cysts. Detection of waterborne protozoan pathogens such as *Giardia* and *Cryptosporidium* have been described in Chapter 6.

An important issue in environmental, microbial-molecular genetics is how various genes are regulated and expressed under various environmental conditions. One of the facts is that some of the environmental microbial pathogens, such as *L. pneumophila* and *V. vulnificus*, alter their gene expression and remain in a dormant stage as nonculturable organisms in the environment. It has also been predicted that several biodegradative microorganisms may not express their degradation genes in the environment. As a result, one may not be sure whether the released GEMs or indigenous microorganisms are degrading the pollutants at a contaminated site. Using specific mRNA as target for PCR amplification and developing a quantitative assay for such method, it is possible to detect the level of mRNA production with high sensitivity in the environmental samples.

A promising method for extraction of specific mRNA from soil seeded with naphthalene-degrading and mercury-resistant bacterial cells has been described [63]. This method can be completed within a few hours, and approximately 17 μg of total RNA per gram (wet weight) of soil containing 8.0×10^8 bacterial cells can be purified with a DNA-RNA hybridization with a detection sensitivity of 160 ng of specific target mRNA. Although this method has the potential for studying in situ gene expression, the humic acid compounds may coprecipitate with samples containing high-cation-exchange-capacity, for example, some sediments that will greatly reduce the total RNA recovery efficiency and sensitivity of detection. Application of PCR for detecting specific mRNA extracted from various environmental samples by this method has yet to be evaluated.

One very important aspect of PCR-gene probe detection of a microbial pathogen in environmental samples is positive amplification signals from nonviable cells providing false-positive results. The ability of the PCR-gene probe methodology to detect boiled or UV-treated, nonviable bacterial cells in water and other environmental samples has been reported [64].

Similarly, biocide-treated, nonviable cells of *S. typhimurium* cells were detected by *Salmonella*-specific primers [65]. Identifying this could be a potential problem in the application and regular monitoring of microbial pathogens or understanding the microbial community structures and interactions among themselves. Thus, targeting messages may be the ideal situation. The use of a rapid and efficient cell-lysis method followed by capturing and purification of total RNA from bacterial cells without degradation of mRNA can be achieved (FastRNA kit, Bio101-Savant).

DETECTION OF AIRBORNE MICROORGANISMS

A number of airborne microbial pathogens are fastidious and it takes a considerable amount of time for their detection by conventional microbiological approach. The use of the PCR-gene probes approach has been applied on the nitrocellulose filters (solid-phase PCR) following capture of a model microorganism *E. coli* DH1 (pWTAla5′) [65a]. This approach has the potential for the detection of airborne microbial pathogens and microorganisms that can not be cultured by conventional microbiological methodologies. Similarly, the detection of airborne varicella-zoster virus (VZV) in hospital rooms was possible using PCR (see Chapter 7). A stepwise protocol for the detection of microorganisms from air samples by using PCR methodologies is described in Chapter 7.

PCR-GENE PROBE HYBRIDIZATION FOR GENETICALLY ENGINEERED MICROORGANISMS

The application of PCR technique for detection and monitoring genetically engineered microorganisms with high sensitivity has been proven. A single copy of the transposon Tn5 was transferred into the genomic DNA of *Rhizobium leguminosarum* released in the soil. These genetically altered microorganisms were detected by "double" PCR amplification using transposon *Tn*5 as the target [17].

The application of PCR techniques for the detection of engineered genes by PCR methodology from the released microorganisms followed by gene probe DNA-DNA hybridization increases the sensitivity of detection of released GEMs in the environment. The application of PCR-gene probe methodology was first applied to monitor genetically engineered microorganisms by amplification of a portion of the 1.3 kb repeat sequence from *Pseudomonas cepacia* AC1100, a herbicide (2,4,5-T)-degrading bacterium, following release in a soil microcosm, to a sensitivity of 100 target cells in 100 grams of sediment against a background of 10^{11} diverse nontarget microorganisms [13]. In a similar study, a 0.3 kb unique DNA sequence from *Pennisetum purpureum* (napier grass) was cloned into pRC10, a derivative of a 2,4-dichlorophenoxyacetic degrading plasmid, and transferred into *Escherichia coli* [67]. This genetically altered microbe was seeded into a filter-sterilized lake and sewage-water samples and detected by PCR at a higher sensitivity than the conventional plating technique even after ten to fourteen days postseeding.

A plant growth–promoting rhizosphere bacterium, *Pseudomonas fluorescens*, was genetically modified by transferring a reporter gene on the chromosome, the mannityl opine catabolism gene (*moc*), from *Agrobacterium tumifaciens* [68]. Nucleic acid–based hybridizations or PCR amplification of the gene was used for the detection of this genetically modified microorganism in the environment.

From these studies it can be concluded that the PCR method can be used for monitoring released GEMs in an environment consisting of a complex habitat of diverse microorganisms where it may be tedious and time-consuming to discriminate the GEMs from the indigenous microorganisms [64].

BIOREMEDIATION STUDIES AND PCR

A number of microorganisms are present in the environment that are involved in the biodegradation of various pollutants and toxic wastes. It is possible to monitor such microorganisms that are involved in bioremediation of the polluted and toxic waste sites by application of the PCR-gene probe methods using conserved regions of the genes that are involved in such function. In a study, using the nucleotide sequence information of a chlorocatechol-degrading gene (*tfd*C) from *Alcaligenes eutrophus* JMP134 (pJP4), it was possible to design oligonucleotide primers for the detection of various chloroaromatic-degrading bacteria by PCR amplification followed by gene-probe hybridization [70]. The PCR amplification followed by gene-probe hybridization using such oligonucleotide primers gives information on the variations, similarities, and functional aspects of various pollutant-degrading genes present in closely or distantly related microorganisms in the environment. The PCR allows the detection of the specific microorganism carrying the degrading gene from a complex mixed population in the environment. When such a polluted site is identified, it is important to investigate the possibility of the presence of various degrading microbes at that site. In addition, by using the mRNA-reverse transcriptase-cDNA-PCR approach, it is possible to determine the degrading activities of these microorganisms at the contaminated sites.

Specific detection of a species of the herbicide (2,4-dichlorophenoxyacetic acid)-degrading bacterium, *A. eutrophus,* was achieved by PCR amplification of a region of *tfd*B gene from pJP4 and its derivative plasmid pRO103 [71]. Using DNA-DNA hybridization, the genetic and phenotypic diversity, and in another study the identification of functionally dominant 2,4-dichlorophenoxyacetic acid–degrading microorganisms in the soil and sediment samples, have been described [72,73]. The PCR amplification of a portion of the *nah* gene followed by Southern blot DNA-DNA hybridization was used for identification of naphthalene-degrading microorganisms in soil and sediment samples [74]. Native bacterial populations were detected from various samples of soil, sediment, and sand by PCR amplification of a conserved region of the 16S rDNA segment and the mercury resistance (*mer*) gene [18]. Since the rDNA target is present in the chromosome and the *mer* gene is located on a plasmid, amplification of both targets simultaneously has the potential to serve as a model system to study the microbial interactions and gene transfer in the natural environment.

APPLICATIONS OF PCR FOR THE DETECTION OF MICROBIAL PATHOGENS IN SEAFOOD AND MEAT PRODUCTS

Foodborne diseases are normally associated with the consumption of contaminated seafood, meat, and various dairy products. Identification and characterization of the etiologic agents of seafood-related disease outbreaks are often impeded by the extended time required to conduct microbiological assays. Protecting public health requires early and rapid diagnosis of pathogens in the food samples. Processing complex food matrices for the purification of PCR-compatible target DNA is a challenging task. Only a few reports exist, and most of them require preenrichment of the sample.

Primarily three different approaches have been made for the purification of target DNA from seafood samples: (1) preenrichment cultivation followed by the purification of DNA by conventional alkaline lysis method. This approach carefully avoids the extensive processing of the food matrices since an aliquot of the preenriched broth is used rather than the entire food sample; (2) direct extraction of total genomic DNAs without any preenrichment step; and (3) following lysis of the contaminated microbial pathogen, the capture of the target DNA using ionic resins or products, such as glass beads.

Microbial pathogens, such as *Vibrio vulnificus, V. cholerae, V. parahaemolyticus,* and *Salmonella typhimurium,* have been detected by PCR, primarily in seafood samples. Also, ample reports are available for the detection of pathogenic *Escherichia coli* in ground beef samples. In milk and other dairy products, the PCR approach has been applied for the detection of *Listeria monocytogenes.* Oligonucleotide primers and other PCR parameters are routine for detection of food-related microbial pathogens. However, purification of the target DNA from complex food samples is still an open area of investigation. In the meantime, preenrichment in the appropriate media is strongly suggested.

ALTERNATE NUCLEIC ACID AMPLIFICATION METHODS AND THEIR POTENTIAL APPLICATIONS IN ENVIRONMENTAL MICROBIOLOGY

Applications of the PCR method in biological sciences, including environmental and medical sciences, generated interest in developing alternative approaches based on the principles of the PCR method, that is, amplification of a defined segment of nucleic acids. Some of these alternative nucleic acid amplification methods are summarized in Table 1.

Among these alternative amplification methods, a single, primer-based "boomerang amplification" method has recently been developed by Kevin Ahern of Oregon State University at Corvallis. In this approach, the genomic DNA is treated with a restriction endonuclease to generate staggered ends to which an adaptor molecule with the panhandle staggered end is complementary to the restriction endonucleases generated site on the genomic DNA. At this stage a single primer is added to the reaction and PCR amplification is performed. A 10,000- to 100,000-fold amplification is achieved within a very short period of time and, unlike conventional PCR approach, the "boomerang PCR" approach requires only one primer.

Another alternative nucleic acids amplification method called the "nucleic acid sequence–based amplification" (NASBA) has been developed for analysis of RNA [75]. The NASBA employs reverse transcriptase, RNAse H, and T7 RNA polymerase, in concert with two oligonucleotide primers, for the amplification of the target RNA in an isothermal condition (Figure 4). The continuous, self-perpetuating repetition of this process results in excess of 1 trillion-fold amplification of the target RNA within ninety minutes. Commercial kits for the detection of food- and waterborne

TABLE 1. In vitro Nucleic Acid Amplification Technology That Is Marketed or Being Marketed by the Selected Biotechnology Companies.

Company	Technology
1. Perkin-Elmer Applied Biosystems Div. (Foster City, CA)	PCR/LDR
2. Cangene Corp. (Mississauga, Ontario, Canada)	Nucleic Acid Sequence-Based Amplification (NASBA)
3. ID Biomedical Corp. (Burnaby, B.C., Canada)	Cycling Probe Technology (CPT)
4. Vysis Inc. (Downers Grove, IL)	Qβ replicase
5. Becton Dickinson Research Center (Research Triangle Park, NC)	Strand Displacement Activation (SDA)
6. Cellmark Diagnostics & Johnson & Johnson (Germantown, MD and Rochester, NY)	Monoclonal-based "hot-start"
7. Incyte Pharmaceutical (Palo Alto, CA)	RNA Amplification Technology
8. Chiron Corp. (Emeryville, CA)	Branched DNA signal amplification

microbial pathogens, such as *Campylobacter, Listeria,* and *Legionella,* are being developed by Cangene Corporation (Mississauga, Ontario, Canada). Also, the NASBA approach has the potential to study *in situ* gene expression of biodegradative microorganisms in toxic waste–contaminated sites. Since NASBA is currently performed under isothermal conditions, nonspecific amplification can be a problem.

Two other alternative methods of nucleic acid amplifications, Qβ replicase [76,77] and ligase chain reaction (LCR) [78], did not gain popularity in microbial diagnostics or environmental sciences. The Qβ replicase approach is quite different from PCR and is based on the mass synthesis of RNA that is complementary to the target DNA sequence. Currently Qβ replicase methodology is relatively complex and would require automation.

The LCR approach was one of the earlier approaches of an alternative method for nucleic acid amplification. The LCR approach utilizes thermostable DNA ligase enzyme. Two (for the sense DNA strand) or four (for both sense and antisense DNA strands) single-stranded, oligonucleotide DNA probes, each approximately 10–20 bases long and complementary to either one or both DNA strands, start the reaction. Following denaturation of the double-stranded target DNA, the oligonucleotide probes are allowed to anneal to the target DNA, lining up end to end and serve as

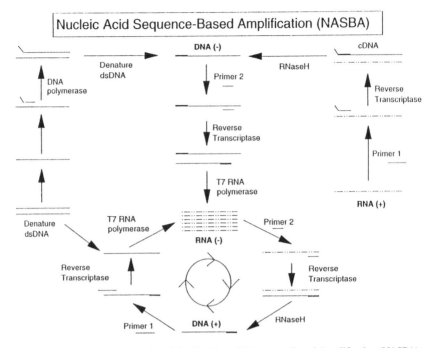

Figure 4 Schematic representation of the Nucleic Acid Sequence-Based Amplification (NASBA) procedure. At first the oligonucleotide primer 1 anneals to the RNA(+) analyte and the cDNA is synthesized using reverse transcriptase. The RNA:cDNA hybrid is treated with RNase H to degrade the RNA strand. The DNA(−) (cDNA) is annealed with oligonucleotide primer 2 and the primer is extended with reverse transcriptase. At this stage, using T7 RNA polymerase, multiple copies of RNA(−) strands are synthesized. Each newly synthesized RNA(−) strand anneals with oligonucleotide primer 2 and DNA(+) strands are made. The newly synthesized DNA(+) strands anneal with oligonucleotide primer 1 and more RNA(−) strands are synthesized using T7 RNA polymerase. Simultaneously, each newly synthesized RNA(−) strand can be used as a template for the synthesis of more DNA(+) strands, which, in turn, can be used for the synthesis of the RNA(−) products [75a].

target for the next cycle. This process is repeated twenty to fifty times yielding million-fold amplifications of the target DNA.

For a short review of these techniques, refer to Lewis [75a].

FUTURE CONSIDERATIONS

The applications of PCR amplification methodologies are very useful in answering and solving many difficult environmentally related biological problems, which have remained unsolved for years due to the limitations of conventional methods. One of the important problems in environmental microbiology is monitoring released GEMs in the environment with high sensitivity. It has been shown that the application of the PCR-gene probe method can detect 1–100 GEMs per gram of soil or sediment, which is a level of sensitivity several orders of magnitude higher than the DNA-DNA hybridization method. Also, in several studies it has been shown that gene-probe hybridization following PCR amplification provides a higher level of detection of the target microorganisms. One of the drawbacks of this approach is that the PCR method requires purified nucleic acid, which can be achieved from the environmental samples through several rigorous methodological steps.

Application of PCR-gene probe technology in monitoring pathogens and indicator microorganisms in water and other environmental samples is very encouraging as a result of the high specificity and sensitivity. The most important criterion in applying this technology to environmental microbiology is the removal of inhibitors and contaminants from the samples. Although a number of inhibitors, including humic acids from various environmental samples, have been identified with possible removal procedures, it is believed that numerous inhibitors may exist which are yet to be identified. For detection of microorganisms in soil and sediments, it has been found that the humic and fulvic acids inhibit polymerase activity and reduce the sensitivity of detection. Although several procedures, such as diluting the samples, ion-exchange chromatography, gel filtration chromatography, PVPP treatment, sucrose gradient purification, and Chelex® 100 (BioRad) resins, have been used to remove humic and fulvic acids and other inhibitors from the samples, none of these methods seems to remove them totally. The application of PCR-gene probe technology to various environmental samples for detection of pathogens and other microorganisms may be affected severely if a relatively universal method for removal of inhibitors from the environmental samples is not developed.

Quantitation of microbial populations by conventional methods has several drawbacks. Although there are several reports on the quantitation of PCR-amplified products by using gene-probe hybridizations, this aspect requires more investigation to develop a reliable quantitation of a microbial population in a given environment. This will permit the study of microbial succession, competition, and community structure in an ecosystem of interest including the microbes living in many extreme environments. As a result of the high efficiency and speed of amplification of the DNA by PCR, it is extremely difficult to estimate the concentration of the original target DNA sequences. However, one of the accepted microbiological techniques, which is traditionally used for quantitation of a microbial population in environmental samples, is the most probable number (MPN) method, which can be used to quantitate the original target DNA of a defined microbial species.

The MPN method is a statistical method that allows for the estimation of microorganisms in a sample. The MPN is based on the Poisson distribution of a microorganism in a solution. A series of tenfold dilutions are inoculated into tubes that contain medium that will give an observable reaction after incubation. Each tube becomes a presence/absence test. This procedure has been successfully applied to PCR as well [79]. Tenfold dilutions are amplified only in those tubes that contain the target sequences. Thus, the presence of amplicons is used to determine the concentration of target sequences based on five- or three-tube series of MPN tables. Although this procedure has a lot of potential, more research is needed to optimize the quantification of microorganisms or target genes in the environment.

Another potential application of PCR-gene probe methodology in the area of environmental microbiology is in distinguishing live from dead cells in a given environmental sample. Although more research needs to be done before applying this approach to actual environmental samples, there have been several reports describing such studies with great promise for the future. However, the concern is the ability of the gene-probe hybridization alone or PCR-gene probe approach to detect nonviable cells in the environment. Recovery of messages and targeting the mRNA has the potential to solve this problem. Alteration of gene expression in many microbial pathogens and other pollutant-degrading microbes due to various environmental conditions is a growing concern to human health; detection of specific mRNA in the environment by PCR-gene probe will give the information of the *in situ* activities of these microorganisms. Much research should be done to target mRNA and optimize the approach in diverse environmental samples.

Arbitrarily primed polymerase chain reaction (AP-PCR) and Randomly Amplified Polymorphic DNA (RAPD) are relatively new, rapid, and simple techniques that generate fingerprints of complex genomes by using single, arbitrarily selected primers to direct amplification where short, single-stranded DNA oligomers are allowed to anneal to denatured template DNA under low stringency conditions [80–83]. At some frequency, two primers will anneal to the template relatively close to one another (100–2000 bases) and onto complementary DNA strands. In the presence of free nucleotides and DNA polymerase, the oligomer primers are extended to form a new copy of the target DNA. By then increasing the stringency of the reaction, only those products formed in the first few low-stringency reactions are amplified. By repeating this process many times, specific fragments of DNA can be amplified and become relatively abundant in the resulting mixture of DNA. Different randomly amplified fragments are generated in different bacteria.

When the amplified DNA fragments are separated based on size by using electrophoresis, they produce a readily distinguishable pattern of DNA
bands. These bands represent a DNA fingerprint that can be used to identify different microbial groups and even various strains within a specific
microbial species. The principles behind AP-PCR are described in Figure
5. Emergence of AP-PCR has special applications in environmental microbiology and solving many of the microbiological activities that have not
been unveiled. The DNA fingerprint analysis will help to trace the specific
microbial strains in an ecosystem and to identify their roles in the ecological activities. Differentiating strains that are virulent from avirulent using
the RNA-AP-PCR approach has the potential to identify the virulent genes
in many of the microorganisms that are found in the environment. The ap-

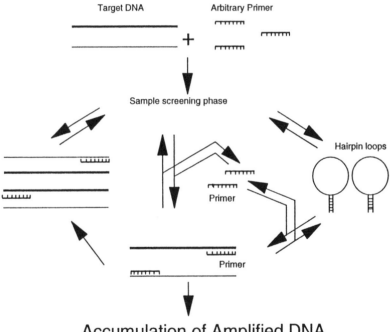

Accumulation of Amplified DNA

Figure 5 A model of interactions between molecular species formed during DNA amplification
with a single, arbitrary oligonucleotide primer. Following the template "screening" phase, a set of
DNA fragments is synthesized. These first-round amplification products are initially single-
stranded and have palindromic termini that allow formation of hairpin loops. In subsequent
rounds of amplification, the products can be in the form of template/template and primer template
duplexes, as well as in single-strand and hairpin loops. The different species produced tend to es-
tablish an equilibrium, while enzyme anchoring and primer extension transform the relatively
rare primer template duplexes into accumulating amplification products.

plication of AP-PCR has great potential for the identification and differentiation of microorganisms. The AP-PCR is a powerful approach to study genomic fingerprints of environmentally important microorganisms. *In situ* PCR is another recent breakthrough in the application of basic PCR methodology originally described by Nuovo et al. [84]. Although, the application of the in situ amplification has not been applied in environmentally related microorganisms, this approach may have the advantage of detecting a single cell in a given environmental sample. Also, the rigorous environmental sample preparation previously discussed may not be necessary for this approach. Simultaneous identification of multiple

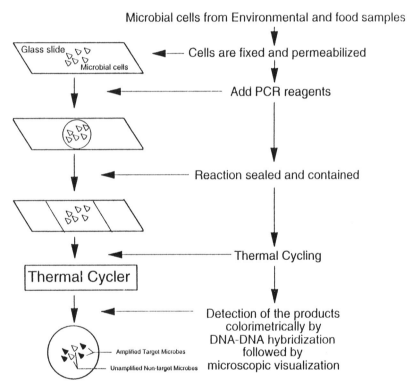

Figure 6 Schematic representation of major procedural steps for in situ PCR DNA amplification in intact microbial cells. Target microbial cells from the environmental or food samples can be recovered following preenrichment (optional) and fixed on a siliconized glass slide. The cells are made permeable by enzymatic treatment, such as pronase, and PCR reagents are added to the cells for RT-PCR. Following permeabilization, total cellular DNAs are destroyed by treatment with DNAse and cDNA is synthesized using reverse transcriptase and an oligonucleotide primer. The PCR amplification reaction is performed in a sealed device and the amplified DNA is detected by fluorescent or colorimetric DNA-DNA hybridization. The positive signals within the cells are visualized by using a fluorescent or a light microscope.

microorganisms in an ecosystem or in an environmental sample by multiplex in situ PCR amplification is a possibility. To study the microbial activities, detection of VBNC state of a microbial pathogen, and differentiation of live versus dead microbial cells, the in situ reverse transcriptase PCR (RT-PCR) approach can be used.

The basic protocol described by Nuovo et al. [84] for *in situ* PCR amplification can be optimized to develop in situ detection of microbial pathogens or microorganisms in various environmental and food samples. Microbial cells in suspension are fixed onto a glass slide. The fixed cells are subjected to PCR amplification using the same parameters for the liquid phase–PCR approach. After addition of the PCR reaction mix, the sample is covered with Amplicover® discs (Perkin-Elmer) and placed into a GeneAmp, *in situ* PCR DNA thermal cycler (Perkin-Elmer or other sources) for the target DNA amplification. The amplified DNA is detected by the incorporation of a fluorescent-labeled deoxynucleotide during amplification or by using fluorescent-labeled specific gene-probe hybridization methods (Figure 6). The RT, *in situ* PCR gene probe DNA-DNA hybridization approach has the potential to study the bacterial gene expression in the environment, such as expression of the toxin genes, biodegradative genes, and genes that are expressed under various environmental stress conditions, avoiding many of the complex sample purifications.

The PCR amplification alone, or coupled with gene probe hybridization, shows promise in cloning genes from environmentally important microorganisms including those organisms that have not yet been cultured. In the near future, technological improvement and subsequent improvement of the PCR methodology with the recent development of various alternative amplification approaches will solve many unanswered questions in microbial ecology and community structure, environmental public health, and molecular analysis of environmental microbiology.

REFERENCES

1 Mullis, K. B. 1990. The unusual origin of the polymerase chain reaction. *Sci. Amer.* 262(4):56–65.

1a Gutell, R. R., N. Larsen and C. R. Woose. 1994. Lessons from an evolving rRNA: 16S and 23S rRNA structures from a comparative perspective. *Microbiol. Reviews.* 58:10–26.

2 Saiki, R. K., D. H. Gelfand, S. Stoffel, S. J. Scharf, R. Higuchi, G. T. Horn, K. B. Mullis, and H. A. Erlich. 1988. Primer-directed enzymatic amplification of DNA with a thermostable DNA polymerase. *Science.* 239:487–494.

3 Cano, R. J., A. Inzunza, D. Norton, E. Bodner. O. Gaber, and F. Tiefenbrunner. 1995. Enzymatic amplification and characterization of fungal DNA from lung tissue of the 5,300 year old Iceman. In: *95th General Meeting of American Society for Microbiology,* Washington D.C., Abstract #F-33.

4 Johnsonbaugh, J. and R. J. Cano. 1995. Isolation of ancient *Bacillus* DNA from sixteen-million-year-old bees in Dominican amber. In: *95th General Meeting of the American Society of Microbiology.* Washington, D.C., Abstract #R-27.

5 Bej, A. K., M. H. Mahbubani, and R. M. Atlas. 1991. Amplification of nucleic acids by polymerase chain reaction (PCR) and other methods and their applications. *Crit. Rev. Biochem. Mol. Biol.* 26(3/4):301-334.

6 Bej, A. K. and M. H. Mahbubani. 1992. Applications of the polymerase chain reaction in environmental microbiology. *PCR Method Appl.* 1:151-159.

7 Bej, A. K. and M. H. Mahbubani. 1994. Applications of the polymerase chain reaction (PCR) in vitro DNA amplification method in environmental microbiology. In: *PCR Technology: Current Innovations,* H. Griffin and A. Griffin (eds.), Academic Press, CA, pp. 327-339.

8 Bej, A. K. and M. H. Mahbubani. 1994. Thermostable DNA Polymerases for in vitro DNA Amplifications. In: *PCR Technology: Current Innovations,* H. Griffin and A. Griffin (eds.), Academic Press, CA, pp. 219-237.

9 Erlich, H. A., D. Gelfand, and J. J. Sninsky. 1991. Recent advances in the polymerase chain reaction. *Science.* 252:1643-1651.

10 Lee, A. B. and T. A. Cooper. 1995. Improved direct PCR screen for bacterial colonies: wooden toothpicks inhibit PCR amplification. *Biotechniques.* 18:225-226.

11 Holben, W. E., J. K. Jansson, B. K. Chelm, and J. M. Tiedje. 1988. DNA probe method for the detection of specific microorganisms in the soil bacterial community. *Appl. Environ. Microbiol.* 54:703-711.

12 Ogram, A., G. S. Sayler, and T. Barkay. 1988. The extraction and purification of microbial DNA from sediments. *J. Microbiol. Meth.* 7:57-66.

13 Steffan, R. J., J. Goksyr, A. K. Bej, and R. M. Atlas. 1988. Recovery of DNA from soils and sediments. *Appl. Environ. Microbiol.* 54:2908-2914.

14 Tsai, Y. and B. H. Olson. 1992. Detection of low numbers of bacterial cells in soils and sediments by polymerase chain reaction. *Appl. Environ. Microbiol.* 58:754-757.

15 Tsai, Y. and B. H. Olson. 1992. Rapid method of separation of bacterial DNA from humic substances in sediments for polymerase chain reaction. *Appl. Environ. Microbiol.* 58:2292-2295.

16 Pillai, S. D., K. L. Josephson, R. L. Bailey, C. P. Gerba, and I. L. Pepper. 1991. Rapid method for processing soil samples for polymerase chain reaction amplification of specific gene sequences. *Appl. Environ. Microbiol.* 57:2283-2286.

17 Young, C. C., R. L. Burghoff, L. G. Keim, V. Minak-Bernero, J. R. Lute, and S. M. Hinton. 1993. Polyvinylpyrrolidone-agarose gel electrophoresis purification of polymerase chain reaction-amplifiable DNA from soils. *Appl. Environ. Microbiol.* 59:1972-1974.

18 Smalla, K., N. Cresswell, L. C. Mendonca, A. Wolters, and J. D. van Elsas. 1993. Rapid DNA extraction protocol from soil for polymerase chain reaction-mediated amplification. *J. Appl. Bacteriol.* 74:78-85.

18a Bruce, K. D., W. D. Hiorns, J. L. Hobman, A. M. Osborn, P. Strike, and D. A. Ritchie. 1992. Amplification of DNA from native populations of soil bacteria by using polymerase chain reaction. *Appl. Environ. Microbiol.* 58:3413-3416.

19 Selenska, S. and W. Klingmuller. 1991. DNA recovery and direct detection of Tn5 sequences from soil. *Lett. Appl. Microbiol.* 13:21–24.

20 Tebbe, C. C. and W. Vahjen. 1993. Interference of humic acids and DNA extracted directly from soil in detection and transformation of recombinant DNA from bacteria and a yeast. *Appl. Environ. Microbiol.* 59:2657–2665.

21 Picard, C., C. Ponsonnet, E. Paget, X. Nesme, and P. Simonet. 1992. Detection and enumeration of bacteria in soil by direct DNA extraction and polymerase chain reaction. *Appl. Environ. Microbiol.* 58:2717–2722.

22 Bachoon, D. and R. E. Hodson. 1995. Electrophoresis purification of sediment DNA. In: *95th General Meeting of the American Society for Microbiology,* Washington, D.C., Abstract #170.

23 Sommersville, C. C., I. T. Knight, R. R. Colwell. 1989. Simple, rapid method for direct isolation of nucleic acids from aquatic environments. *Appl. Environ. Microbiol.* 55:548–554.

24 Paul, J. H., L. Cazares, and J. Thurmond. 1990. Amplification of the *rbc*L gene from dissolved and particulate DNA from aquatic environments. *Appl. Environ. Microbiol.* 56:1963–1966.

25 Weller, R. and D. M. Ward. 1989. Selective recovery of 16S rRNA sequences from natural microbial communities in the form of cDNA. *Appl. Environ. Microbiol.* 55:1818–1822.

26 Fuhrman, J. A., D. E. Comeau, A. Hagstrom, and A. M. Chan. 1988. Extraction from natural planktonic microorganisms of DNA suitable for molecular biological studies. *Appl. Environ. Microbiol.* 54:1426–1429.

27 Lee, S. and J. A. Fuhrman. 1990. DNA hybridization to compare species compositions of natural bacterioplankton assemblages. *Appl. Environ. Microbiol.* 56:739–746.

28 Zehr, J. P. and L. A. McReynold. 1989. Use of degenerate oligonucleotides for the amplification of the *nif*H gene from the marine cyanobacterium *Trichodesmium thiebautii. Appl. Environ. Microbiol.* 55:2522–2526.

29 Bej, A. K., M. H. Mahbubani, J. L. DiCesare, and R. M. Atlas. 1991. PCR-gene probe detection of microorganisms using filter-concentrated samples. *Appl. Environ. Microbiol.* 57:3529–3534.

30 Oyofo, B. A. and D. M. Rollins. 1993. Efficacy of filter types for detecting *Campylobacter jejuni* and *Campylobacter coli* in environmental water samples by polymerase chain reaction. *Appl. Environ. Microbiol.* 59:4090–4095.

31 Bej, A. K., R. J. Steffan, J. L. DiCesare, L. Haff, and R. M. Atlas. 1990. Detection of coliform bacteria in water by polymerase chain reaction and gene probes. *Appl. Environ. Microbiol.* 56:307–314.

32 Bej, A. K., M. H. Mahbubani, R. Miller, J. DiCesare, L. Haff, and R. M. Atlas. 1990. Multiplex PCR amplification and immobilized capture probe for detection of bacterial pathogens and indicators in water. *Mol. Cell. Prob.* 4:353–365.

33 Boccuzzi, V. M., W. L. Straube, J. Ravel, R. R. Colwell, and R. T. Hill. 1995. Removal of contaminating substances from environmental samples prior to PCR by using Sephadex G-200 spun columns. In: *95th General Meeting for American Society for Microbiology,* Washington, D.C. Abstract #N-168.

34 Olson, B. H. 1991. Tracking and using genes in the environment. *Environ. Sci. Technol.* 25:604–611.

35 Hofle, M. G. 1989. *Recent Advances in Microbial Ecology.* Japan Scientific Press, Tokyo, p. 1239.

36 Moran, M. A., V. L. Torsvik, T. Torsvik, and R. E. Hodson. 1993. Direct extraction and purification of rRNA for ecological studies. *Appl. Environ. Microbiol.* 59:915–918.

37 Jeffrey, W. H., S. Nazaret, and R. Von-Haven. 1994. Improved method for recovery of mRNA from aquatic samples and its application to detection of *mer* expression. *Appl. Environ. Microbiol.* 60:1814–1821.

38 Craun, G. F. 1988. Surface water supplies and health. *J. Amer. Water Works Assoc.* 80:40–52.

39 DeLeon, R., Shieh, C., Baric, R. S., and Sobsey, M. D. 1990. Detection of enteroviruses and hepatitis A virus in environmental samples by gene probes and polymerase chain reaction. In: *Proceedings of the 1990 Water Quality Technology Conference.* American Water Works Association, Denver, pp. 833–853.

40 Ansari, S. A., S. R. Farrah, and G. R. Chaudhry. 1992. Presence of human immunodeficiency virus nucleic acids in wastewater and their detection by polymerase chain reaction. *Appl. Environ. Microbiol.* 58:3984–3990.

41 Abbaszadegan, M., M. S. Huber, C. P. Gerba, and I. L. Pepper. 1993. Detection of enteroviruses in groundwater with the polymerase chain reaction. *Appl. Environ. Microbiol.* 59:1318–1324.

42 Tsai, Y. L., B. Tran, L. R. Sangermano, and C. J. Palmer. 1994. Detection of poliovirus, hepatitis A virus, and rotavirus from sewage and ocean water by triplex reverse transcriptase. *Appl. Environ. Microbiol.* 60:2400–2407.

43 Deng, M. Y., S. P. Day, and D. O. Cliver. 1994. Detection of hepatitis A virus in environmental samples by antigen-capture PCR. *Appl. Environ. Microbiol.* 60:1927–1933.

44 Starnbach, M. N., S. Falkow, and L. S. Tompkins. 1990. Species-specific detection of *Legionella pneumophila* in water by DNA amplification and hybridization. *J. Clin. Microbiol.* 27:1257–1261.

45 Mahbubani, M. H., A. K. Bej, R. Miller, L. Haff, J. DiCesare, and R. M. Atlas. 1990. Detection of *Legionella* with polymerase chain reaction and gene probe methods. *Mol. Cell. Probes.* 4:175–187.

46 Jones, D. D., R. Law, and A. K. Bej. 1993. Detection of *Salmonella* spp. in contaminated oysters using polymerase chain reaction (PCR) and gene probes. *J. Food Sci.* 58:1191–1197.

47 Bej, A. K. and M. H. Mahbubani, M. J. Boyce, and R. M. Atlas. 1994. Detection of *Salmonella* in shellfish by using polymerase chain reaction. *Appl. Environ. Microbiol.* 60:368–373.

48 Mahbubani, M. H., A. K. Bej, M. H. Perlin, F. W. Schaeffer III, W. Jakubowski, and R. M. Atlas. 1992. The differentiation of *Giardia duodenalis* from other *Giardia* spp. based on the polymerase chain reaction and gene probes. *J. Clin. Microbiol.* 30:74–78.

49 Mahbubani, M. H., A. K. Bej, M. H. Perlin, F. W. Schaeffer III, W. Jakubowski, and R. M. Atlas. 1991. Detection of *Giardia* using the polymerase chain reaction and distinguishing live from dead cysts. *Appl. Environ. Microbiol.* 57:3456–3461.

50 Bej, A. K., J. L. DiCesare, L. Haff, and R. M. Atlas. 1991. Detection of *Escherichia coli* and *Shigella* spp. in water by using polymerase chain reaction (PCR) and gene probes for *uid. Appl. Environ. Microbiol.* 57:1013–1017.

51 Bej, A. K., S. C. McCarty, and R. M. Atlas. 1991. Detection of coliform bacteria and *Escherichia coli* by multiplex polymerase chain reaction: Comparison with defined substrate and plating methods for water quality monitoring. *Appl. Environ. Microbiol.* 57:2429–2432.

52 Kadokami, Y. and R. V. Lewis. 1990. Membrane bound PCR. *Nucleic Acids Res.* 18:3082.

52a Cleuziat, P. and J. Baudouy-Robert. 1991. Specific detection of *Escherichia coli* and *Shigella* species using fragments of genes coding for beta-glucuronidase. *FEMS Microbiol. Letts.* 72:315–322.

53 Toranzos, G. A. and A. J. Alvarez. 1991. Solid-phase polymerase chain reaction: Applications for direct detection of enteric pathogens in waters. *Can. J. Microbiol.* 38:365–369.

54 Chamberlain, J. S., R. A. Gibbs, J. E. Ranier, P. N. Nguyen, and A. Radolf. 1988. Deletion screening of Duchenne muscular dystrophy locus via multiplex DNA amplification. *Nuc. Acid. Res.* 16:11141–11156.

55 Bej, A. K., M. H. Mahbubani, and R. M. Atlas. 1991. Detection of viable *Legionella pneumophila* by using polymerase chain reaction and gene probes. *Appl. Environ. Microbiol.* 57:597–601.

55a Lang, A. L., Y. L. Tsai, C. L. Mayer, K. C. Patton, and C. J. Palmer. 1994. Multiplex PCR for detection of the heat-labile toxin gene and shiga-like toxin I and II genes in *Escherichia coli* isolated from natural waters. *Appl. Environ. Microbiol.* 60:3145–3149.

56 Bej, A. K., M. H. Mahbubani, and R. M. Atlas. 1991. Detection of viable *Legionella pneumophila* in water using PCR and gene probe methods. *Appl. Environ. Microbiol.* 57:597–600.

56a Beasley, L., D. D. Jones, and A. K. Bej. 1994. A rapid method for detection and differentiation of KP+ and KP− *Vibrio parahemolyticus* in artificially contaminated shellfish by in vitro DNA amplification and gene probe hybridization methods. In: *94th General Meeting of American Society for Microbiology (ASM)*, Las Vegas, Nevada.

57 Rasmussen, S. R., H. B. Rasmussen, M. R. Larsen, R. Hoff-Jorgensen, and R. J. Cano. 1994. Combined polymerase chain reaction-hybridization microplate assay used to detect bovine leukemia virus and *Salmonella*. *Clin. Chem.* 40(2):200–205.

58 Brauns, L. A., M. C. Hudson, J. D. Oliver. 1991. Use of the polymerase chain reaction in detection of culturable and nonculturable *Vibrio vulnificus* cells. *Appl. Environ. Microbiol.* 57:2651–2655.

59 Colwell, R. R., P. R. Brayton, D. J. Grimes, D. B. Roszak, S. A. Huq, and L. M. Palmer. 1985. Viable but nonculturable *Vibrio cholerae* and released pathogens in the environment: Implication for release of genetically engineered microorganisms. *Bio/Technology.* 3:817–820.

60 Hussong, D., R. R. Colwell, M. O. O'Brien, E. Weiss, A. D. Pearson, R. M. Eiener, and W. D. Burge. 1987. Viable *Legionella pneumophila* not detectable by culture on agar media. *Bio/Technology.* 5:947–950.

61 Mahbubani, M., A. K. Bej, J. DiCesare, R. Miller, L. Haff, and R. M. Atlas. 1991. Detection of bacterial mRNA using polymerase chain reaction. *Bio/Techniques.* 10:48–49.

62 Roszak, D. B. and Colwell, R. R. 1987. Survival strategies of bacteria in the natural environment. *Microbiol. Rev.* 51:365–379.

63 Tsai, Y., M. J. Park, and B. H. Olson. 1991. Rapid method for direct extraction of mRNA from seeded soils. *Appl. Environ. Microbiol.* 57:765–768.

64 Josephson, K. L., C. P. Gerba and I. L. Pepper. 1993. Polymerase chain reaction detection of nonviable bacterial pathogens. *Appl. Environ. Microbiol.* 59: 3313–3315.

65 Graves, S. and A. K. Bej. 1994. Use of polymerase chain reaction (PCR) in distinguishing live *Salmonella typhimurium* from biocide-treated dead cells in water. In: *94th General Meeting of the American Society for Microbiology (ASM)*, Las Vegas, Nevada.

65a Alvarez, A. J., M. P. Buttner, G. A. Toranzos, E. A. Dvorsky, A. Toro, T. B. Heikes, L. E. Mertikas-Pifer, and L. D. Stetzenbach. 1994. Use of solid-phase PCR for enhanced detection of airborne microorganisms. *Appl. Environ. Microbiol.* 60:374–376.

66 Sawyer, M. H., Chamberlin, C. J., Wu, Y. N., Aintablian, N. and M. R. Wallace. 1994. Detection of varicella-zoster virus DNA in air samples from hospital rooms. *J. Infect. Dis.* 169:91–94.

67 Chaudhry, G. R., G. A. Toranzos, A. R. Bhatti. 1989. Novel method for monitoring genetically engineered microorganisms in the environment. *Appl. Environ. Microbiol.* 55:1301–1304.

68 Hwang, I. and S. K. Farrand. 1994. A novel gene tag for identifying microorganisms released into the environment. *Appl. Environ. Microbiol.* 60:913–920.

69 Molin, S., L. Boe, L. B. Jensen, C. S. Kristensen, M. Givskov, J. L. Ramos, and A. K. Bej. 1993. Suicidal genetic elements and their use in biological containment of bacteria. *Ann. Rev. Microbiol.* 47:139–166.

70 Greer, C. W., D. Beaumier, H. Bergeron, and P. C. K. Lau. 1991. Polymerase chain reaction isolation of a chlorocatechol dioxygenase gene from a dichlorobenzoic acid degrading *Alcaligenes denitrificans*. In: *91st General Meeting of the American Society for Microbiology.* Abstract Q-99:292.

71 Neilson, J. W., K. L. Josephson, S. D. Pillai, and I. L. Pepper. 1992. Polymerase chain reaction and gene probe detection of the 2,4-dichlorophenoxyacetic acid degradation plasmid pJP4. *Appl. Environ. Microbiol.* 58:1271–1275.

72 Ka, J. O., W. E. Holben and J. M. Tiedje. 1994. Use of gene probes to aid in recovery and identification of functionally dominant 2,4-dichlorophenoxyacetic acid-degrading populations in soil. *Appl. Environ. Microbiol.* 60:1116–1120.

73 Ka, J. O., W. E. Holben, and J. M. Tiedje. 1994. Genetic and phenotypic diversity of 2,4-dichlorophenoxyacetic acid (2,4-D)-degrading bacteria isolated from 2,4-D treated field soils. *Appl. Environ. Microbiol.* 60:1106–1115.

74 Herrick, J. B., E. L. Madsen, C. A. Batt, and W. C. Ghiorse. 1993. Polymerase chain reaction amplification of naphthalene-catabolic and 16S rRNA gene sequences from indigenous sediment bacteria. *Appl. Environ. Microbiol.* 59:687–694.

75 Sooknanan, R. and L. T. Malek. 1995. NASBA: A detection and amplification system uniquely suited for RNA. *Bio-Technology.* 13:563–564.

75a Lewis, R. 1995. Rival amplification techniques shake-up reveals a bioindustrial segment in flux. *Gen. Eng. News.* 15:1–25.

76 Lizardi, P. M., C. E. Guerra, H. Lomeli, I. Tussie-Luna, and F. R. Kramer. 1988. Exponential amplification of recombinant-RNA hybridization probes. *Bio-Technology.* 6:1197–1201.

77 Knight, P. 1989. Amplifying probe assays with Q-beta replicase. *Bio-Technology.* 7:609–611.

78 Barany, F. 1991. Genetic disease detection and DNA amplification using cloned thermostable ligase. *Proc. Natl. Acad. Sci. USA.* 88:189–193.

79 Toranzos, G. A., A. J. Alvarez and E. A. Dvorsky. 1992. Application of the polymerase chain reaction technique to the detection of pathogens in water. *Wat. Sci. Tech.* 27:207–210.

80 Caetano-Anolles, G., B. J. Bassam, and P. M. Gresshoff. 1992. Primer-template interaction during DNA amplification fingerprinting with single arbitrary oligonucleotides. *Mol. Gen. Genet.* 235:157.

81 Caetano-Anolles, G. 1993. Amplifying DNA with arbitrary oligonucleotide primers. *PCR Meth & Appl.* 3:85.

82 Welsh, J. and M. McClelland. 1990. Fingerprinting genomes using PCR with arbitrary primers. *Nucl. Acid. Res.* 18:7213.

83 Welsh, J. and M. McClelland. 1991. Genomic fingerprinting using arbitrarily primed PCR and a matrix of pairwise combinations of primers. *Nucl. Acid. Res.* 19:5275.

84 Nuovo, G. J., P. MacConnell, and F. Gallery. 1994. Analysis of non-specific DNA synthesis during *in situ* PCR and solution-phase PCR. *PCR Meth. Appl.* 4:89–96.

CHAPTER 2

Basic Methodology for DNA and RNA Amplification

ABDIEL J. ALVAREZ[1]
GARY A. TORANZOS[2]

INTRODUCTION

THE development of methods capable of detecting low concentrations of microorganisms is of paramount importance in view of the low infectious dose of many pathogens [1]. Exponential amplification of a target sequence using the Polymerase Chain Reaction (PCR) technique can attain the sensitivity and specificity required in environmental studies. Of the basic molecular biology techniques developed in the last decade, none has had a greater impact than PCR [2].

DNA amplification using the PCR technique has recently been used in various environmental studies to detect microorganisms in drinking water, soil, and air samples. *Cryptosporidium, Giardia,* enterovirus, *Entamoeba, Escherichia coli,* other environmentally important bacteria such as *Aeromonas* and *Legionella,* and various genetically engineered microorganisms have been detected [3–12]. The greatest impact of the PCR has been on the detection of microorganisms for which *in vitro* cultivation is lengthy, inconvenient, expensive, or simply unavailable.

A typical PCR amplification includes three main steps: denaturation, annealing, and polymerization (Figure 1). The method involves separating the DNA strands and annealing short, specific, oligonucleotide primers to regions flanking or within a target sequence. The DNA polymerase is

[1]Harry Reid Center for Environmental Studies, University of Nevada-Las Vegas, Box 454009, Las Vegas, NV 89154-4009, U.S.A. (Current address: Abbott Diagnostics, Inc., P.O. Box 278, Barceloneta, Puerto Rico, 00617.)
[2]Department of Biology, P.O. Box 23360, The University of Puerto Rico, Rio Piedras, Puerto Rico 00931-3360.

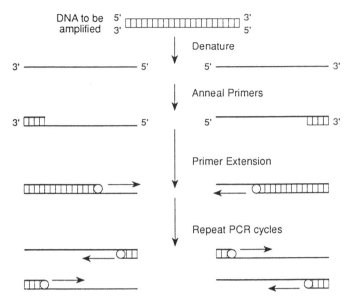

Figure 1 Diagram of the three main steps of a PCR amplification of target DNA.

added to the reaction mix in the presence of deoxynucleotides, and DNA is polymerized from the primers across the target region. The new, double-stranded DNA (dsDNA) is again denatured and the process is repeated for several cycles. This results in a theoretical exponential accumulation of the specific target 2^n, where n is the number of cycles of denaturation and polymerization [13]. In practice, amplification efficiency drops in response to a number of factors; among them is a limited concentration of reagents as the reaction proceeds. However, it has been shown that a single copy of a genomic sequence can be amplified by a factor of more than 10 million with high specificity [14]. The method has been improved by the use of thermostable DNA polymerases, which eliminate the need to add new polymerase after each cycle, and the introduction of thermal cyclers that enhance the reliability of the technique.

Applications of the PCR technique are not limited to the amplification of a specific target. Several strategies for applying it to a variety of research problems have been described. Research applications have sometimes helped to overcome one of the requirements of the PCR, which is the need for information on the sequence of the target to design the primers (e.g., inverse PCR [15]). Other variations include the introduction of new sequences, such as restriction sites, regulatory elements or mutations into the target [16], quantitative PCR [17], sequencing [18], and multiplex PCR [19,20]. A detailed presentation of all relevant procedures in which the PCR technique has been used is beyond the scope of this chapter. Thus, the goal is to present the basics of the PCR and a strategy for a start-up amplification of a given sequence. The chapter will concentrate on the parameters of the reaction and factors to consider for a successful amplification with the highest yield of product and/or the lowest yield of non-specific amplification products.

AMPLIFICATION CONDITIONS

The incorporation of thermostable DNA polymerase into the amplification process was a significant improvement to the reaction [14]. However, precision in amplification parameters was required in order to avoid minute variations in the reaction that could result in unsuccessful experiments. With the introduction of an automated thermal cycler, most of the deficiencies associated with manual amplification were eliminated. This instrument effectively automated PCR by providing accurate control of heating and cooling cycles. In addition to high precision, perhaps one of the greatest advantages of automated PCR is the rapidity with which amplified products are obtained. A typical run in most instruments is twenty-five to thirty-five cycles and takes approximately two to four hours for completion, depending on the segment times used.

The three main steps of the PCR are (1) denaturation, (2) annealing, and (3) polymerization. Specific reaction conditions are determined primarily based on the primers chosen. The thermal cycler can be programmed with different "soak" and "ramp" times for each step (Figure 2). Soak time is the amount of time at which the temperature remains the same at a given step. The size of the fragment being amplified is the critical factor in determining an appropriate soak time. As a starting point, allow one minute for each kilobase of the desired PCR product. Although this is almost cer-

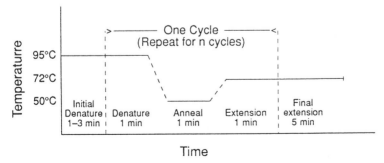

Figure 2 Diagram of a typical PCR amplification cycle profile.

tainly excessive, it is a good place to begin. Ramping time, the rate of change in temperature, can be critical for some applications of the technique that require a gradual change in temperature between steps. More often a ramp time with maximum speed between steps is used.

For maximum amplification efficiency, full denaturation of the target DNA is needed before the start of cycling. This step can be done without the thermostable polymerase. When working with environmental samples, this step can inactivate endogenous enzymes before adding the polymerase. For denaturation a temperature between 94°C and 96°C is usually sufficient for complete separation of the double-stranded DNA. During this step it is important to remember that some of the activity of the DNA polymerase may be lost at higher temperatures; thus, only the necessary exposure of the enzyme to this temperature is recommended.

Annealing temperature is primarily determined by the $G + C$ content of the primers. A range from 5°C to 25°C below the melting temperature (T_m) is generally recommended, and a commonly used temperature is 55°C. Amplifications performed below this temperature or with a lengthy soak may result in the amplification of nonspecific products, which can be seen as false positives, or the presence of more than one amplicon in a gel electrophoresis of the products. A high annealing temperature increases the stringency and probably the specificity of the reaction. However, this may reduce the efficiency of the reaction. Because of the large molar excess of primers usually present in the reaction mix, annealing occurs almost instantaneously, and a long incubation step is not required.

The third step, extension, is carried out at the optimum temperature range for polymerization of the thermostable enzyme used. For most enzymes a 70–74°C range is recommended. Although primer extension is commonly performed at a higher temperature than that of the annealing step, there is no theoretical justification for this practice. Annealing and extension at a single temperature have been shown to be equally effective because of the high processivity of Taq polymerase [21]. In most amplifications a final extension step (a final soak at 72°C) to obtain fully double-stranded molecules is performed.

Incubation times at each step and the number of cycles should be kept to a minimum in order to reduce the possibility of nonspecific contaminant amplification. The appropriate number of cycles is related to the abundance of target DNA and efficiency of the PCR. It is generally safe to assume that if more than forty cycles are required to detect the target sequence, then reaction conditions are suboptimal. In addition, false positives may occur as a result of the presence of extremely small concentrations of sequences similar to those of the target. The fewer the number of PCR cycles used, the "cleaner" the results (usually shown as a single, sharp amplicon band after gel electrophoresis).

After thirty-five cycles a plateau of amplification is usually observed. This effect may be caused by a combination of factors, including competition by nonspecific products, substrate excess, and product reassociation [22]. In the first case, the unwanted DNA fragments compete with the desired fragment for the binding sites of the limiting reagents (i.e., polymerase, primers). Substrate excess is the result of having synthesized more DNA than the amount of polymerase present in the reaction is capable of replicating in the given extension time. Finally, end product inhibition may be caused by reassociation of the single-stranded PCR products before the annealed primers can be extended or by incomplete denaturation. Usually these problems may be overcome by extending the soak times of each step as cycles progress.

Sometimes direct amplification of a specific target DNA is not successful, which may be due to only a few copies of the target sequence present or because of the presence of organic and inorganic compounds that may be partially inhibitory to the DNA polymerase. Ruano et al. described a diphasic amplification strategy, termed "booster PCR," to improve the efficiency of amplifying target sequences present in low numbers [23]. Taking a small aliquot of the reaction mixture that has already been amplified and placing it in a fresh reaction mixture results in an additional exponential amplification as well as diluting any inhibitory compounds that may be present [11,23]. This inhibitory phenomenon has also been observed in clinical samples [24].

An alternative to the booster PCR is the use of nested primers [25]. After amplification of a DNA target using a primer set, an aliquot is amplified a second time using an inner (nested) primer set. The nested amplification has been shown to be more successful than diluting and reamplifying with the same primers [26].

Primer-dimers and nonspecific products are often obtained along with the specific amplification product due to the permissive conditions for annealing/extension that may occur at room temperature. One way to reduce these artifacts, and at the same time increase the efficiency and quality of the PCR reaction, is to perform a "hot start" amplification [27]. To hot start the PCR, the reagents are combined except for either the dNTPs or primers and preheated to about 80°C for two to three minutes. The remaining components are then quickly added, overlaid with mineral oil, and amplified. The addition of *E. coli* ssDNA-binding protein has also been reported to increase specific amplification [28]. The use of Ampliwax® PCR (Perkin-Elmer) facilitates the hot start technique and replaces mineral oil as a vapor barrier. The Ampliwax® prevents a mispriming at room temperature by creating a solid wax layer above a reagent mix that lacks a key reagent. The omitted essential reagent is then added to the top of the wax layer. During the first denaturation step the wax layer melts allowing all the reagents to mix thoroughly while the wax layer rises to the surface and creates the barrier that prevents evaporation. The hot start and Ampliwax®, in addition to enhancing PCR uniformity, are especially useful in applications with degenerate primers and multiplex PCR.

The development of a programmable temperature cycler has greatly simplified and improved the PCR amplification process. Several manufacturers have developed thermal cyclers where the temperature, incubation time, and cycle numbers can be programmed. While the performance of many machines is somewhat variable, most can be used to perform DNA amplification. Cost and service are the major factors to consider when selecting cyclers. The reaction vessel for most machines is a 0.5-ml tube that fits in a metal block designated to hold the tubes, but some manufacturers produce modified blocks that hold 96-well microtiter plates, slides, and smaller tubes. To increase the rate of temperature transfer, thin-walled microtubes and capillary tubes are being used in some applications. It has been reported that variations in temperature across the heat block can affect the amplification [29]. To monitor the homogeneity of cooling and heating across the whole block, a microprobe thermocouple connected to a recording device, which shows the actual temperature inside the reaction tubes during each run, is recommended. Some manufacturers supply these with the machines, while other machines can be adapted.

Also available is a heated-lid accessory for the thermal cyclers that prevents evaporation and allows for oil-free operation. Other manufacturers have avoided using a metal block and have adapted the equipment to be used in an oven, which assures an even temperature in all reaction vessels. However, they are slower to reach the temperatures between steps. As previously mentioned, capillary tubes can be used as reaction vessels in commercially available "rapid-air" cyclers, which can allow the reaction to be completed in twenty minutes. The limitation of this equipment is a small reaction volume.

STANDARD PROTOCOL FOR DNA AMPLIFICATION

In environmental samples where the template DNA is likely to be the most important variable in a PCR reaction, it is impossible to control the quality as well as the precise concentration. In addition, the DNA varies in GC content, possible chemical modifications, tertiary structures, or other parameters that can make the sample difficult to amplify. There are no specific guidelines on what buffer conditions should be used for a particular template and primer set.

Optimization of the PCR amplification for a given application is highly recommended. These "standard" conditions should then allow for amplification of the target sequences. It is important to note that the efficiency and specificity of the technique depend on a number of variables, which are affected to different degrees by such factors as primer sequence, thermal cycling parameters, and the number of cycles. Each component of the reaction mix for each primer set and target DNA should be optimized. Sometimes it is necessary to adjust the reaction if the source of any component of the reaction is changed. Therefore, the following so-called standard conditions should be regarded as a point of departure to explore modifications and potential improvements. The following is a general protocol for DNA amplification.

PROTOCOL I

(1) Set up a 100-μl reaction in a 0.5 ml microfuge tube containing:
 • 10 μl 10X PCR reaction buffer [10 mM Tris-HCl (pH 8.4), 1.5 mM MgCl$_2$, 50 mM KCl, 0.01% (v/v) Tween 20, 0.01% (w/v) gelatin]
 • 2 μl dNTP mixture (10 mM each dNTP)
 • 2 μl 5′ primer (20 μM)
 • 2 μl 3′ primer (20 μM)
 • 73.5 μl dH$_2$O
 • 10 μl Template DNA ($\geq$ 10^4 targets)
 • 0.5 μl (2.5 U) of Taq DNA polymerase
(2) Mix, centrifuge briefly, and overlay with mineral oil (40–80 μl).
(3) Denature the dsDNA at 96°C for two minutes.
(4) Perform twenty-five to thirty-five cycles of amplification using the following parameters:
 • denaturation: 96°C, one minute
 • annealing: 50°C, one minute
 • extension: 72°C, two minutes
(5) Conclude with a final extension at 72°C for five minutes.
(6) Store amplified samples at −20°C until analysis.

REACTION COMPONENTS

DNA POLYMERASE

Before the discovery of thermostable DNA polymerases, DNA amplification was accomplished using the DNA polymerase Klenow fragment [6,30]. This process is a common DNA synthesis, the difference being that a defined sequence is copied repeatedly resulting in an exponential amplification. The addition of fresh enzyme after each denaturation step was required, resulting in high reaction costs, and a tedious procedure. The availability of thermostable Taq polymerase in early 1987 resulted in a significant improvement and simplification of the amplification process [14]. The use of a thermostable polymerase allows for the use of high annealing and extension temperatures during the amplification cycling, which results in fewer nonspecific amplification products. In addition to the increase in specificity and yield, use of Taq allows the amplification of much longer fragments than does Klenow enzyme.

Thermostable DNA polymerases (Taq being the most commonly used) are available from many vendors. We have found no significant differences in the performance of thermostable polymerases from different suppliers in typical amplification protocols if appropriate buffer conditions are used. The enzymes are available in their native forms, but further improvements to PCR reproducibility came when the enzyme was cloned and expressed in different bacterial hosts. The half-life of Taq polymerase is approximately forty minutes at $95°C$ [31]. This provides sufficient thermostability so that there is no substantial loss of enzymatic activity in most PCR protocols. Under optimal conditions this enzyme can extend a new DNA strand at a rate of up to 150 nucleotides/sec [31], thus, for most amplification targets a thirty- to sixty-second extension step is sufficient. In a standard PCR amplification 1–2.5 units of Taq per 100 μl reaction is used. Too much enzyme can lead to nonspecific background, which appears as smearing on agarose gels, and too little may result in low yield of the desired amplification product.

It has been shown that different DNA polymerases have distinct characteristics, which affect the efficacy of PCR. Recent developments in enzyme technology have introduced new thermostable polymerases with proofreading ability that Taq lacks, resulting in a lower miscorporation rate. Also, a truncated Taq polymerase, the Stoffel fragment [32], is available for amplification protocols. This enzyme is more thermostable (by twofold at $97.5°C$) than Taq, allowing denaturation steps for longer periods or at higher temperature, which can help avoid problems with the secondary structure of some amplification targets. The Stoffel fragment lacks any $5'-3'$ exonuclease activity and also exhibits optimal activity over a broader range of Mg^{++} concentration, which is especially useful for applications such as multiplex PCR. The UlTma DNA polymerase (Perkin-Elmer) is one of the most recent thermostable polymerases available. This is the first polymerase specifically designed for its ability to repair $3'$ mismatches in PCR amplification, to provide a high yield of a specific PCR product, and produce blunt-ended products suitable for subsequent cloning.

PRIMERS

The specificity of PCR is the result of the selection of synthetic oligo-nucleotide primers to anneal to a unique sequence in the template. Selection of good primers increases the chances of success. Identifying sequences suitable for priming DNA synthesis is the initial step in performing the PCR. A database analysis of the target sequence against other known sequences of the organism or any other related organisms (i.e., standard computer-sequence homology program) is recommended to avoid unwanted annealing, which can result in false positive results. Major considerations for primer selection are discussed below.

A primer should have sufficient length to provide specific priming but be short enough to allow rapid annealing and thus form a stable, double-stranded template to which the polymerase can bind. While the recommended primer length is 18–28 bases, longer primers are suggested for the detection of sequences that are not well conserved. This is because a single base change is less likely to affect annealing as primer length increases. It is important that primers have compatible melting temperatures, an appropriate range being 50–60°C. Melting temperature (T_m) is defined as the dissociation temperature of the primer-template duplex, and a rough approximation for calculating T_m is 2°C for each A-T and 4°C for each G-C in the primer sequence. Also, the G-C base composition, which helps formation of stable hybrids, should be 40–60% in both primers. Since primer extension occurs starting from the 3′ end, the single 3′ base of the primer should be G or C. This stabilizes annealing and promotes the polymerase activity [33]. Extra sequence information not complementary to the template can be added to the 5′ end of the primer without an effect on annealing. These sequences can serve to introduce restriction sites or regulatory elements (i.e., T7 RNA polymerase-binding site, GC clamp) at the ends of amplified target sequences.

It is very important to check for possible complementarity of all the primers added in the reaction. Interstrand complementary regions at the 3′ ends of the primers should be avoided to limit primer-dimer formation, which can interfere in the reaction. This is critical in multiplex amplification since more than one set of primers is present. The presence of palindromic and intrastrand complementary regions should also be avoided to prevent primer-template annealing problems due to secondary structure of the primer. To reduce the possibility of nonspecific amplification, primers with significant degrees of complementarity to known repeated elements should not be used.

Primers can be obtained from commercial suppliers or in-house facilities. The HPLC-purified primers may give better results. However, highly purified primers are not required for most PCR protocols. The integrity and concentration of the primers should be verified by gel electrophoresis and spectrophotometrically as soon as they are received. Primers should be aliquoted and kept at $-20°C$. Working concentration of each primer should be between 0.1 μM and 0.5 μM (final) and equimolar quantities of each one must be used. High primer concentration may promote misprim-ing that can generate nonspecific product and primer-dimer formation. Primer-dimers are dsDNA PCR products consisting of the two primers and their complementary sequences, but additional bases may be inserted between the primers [27]. These products may compete with the target DNA for primer, dNTPs and enzyme, thus interfering with the efficiency of the amplification.

Although the manual selection of optimal primers sets is possible, it can be quite tedious. Computer software available through Internet or com-mercial suppliers can test thousands of primer combinations from any DNA/RNA sequence in seconds. They are easy to use and allow for the selection of not only unique but highly specific primers that are free of se-quence repeats (dimers or hairpins). Most programs calculate the optimal annealing temperature, PCR product composition, accurate T_m, and ex-tinction coefficients. The use of computer software allows one to create standardized criteria for primer selection within the laboratory, always leaving the opportunity to combine any pair of primers (i.e., multiplex, nested, seminested).

DEOXYNUCLEOTIDE TRIPHOSPHATES (dNTPs)

The most important aspect concerning the dNTPs is that each nucleotide should be used at equal concentration to minimize misincorporation, which increases the error rate by influencing nucleotide selectivity at the insertion step. One must ensure that the reaction mixture contains nonlimiting (i.e., excess) amounts of dNTPs, although too high a concen-tration tends to promote misincorporations. A titration in the dNTP con-centration in a reaction leads to improvements in fidelity. Successful PCR amplifications can be achieved with concentrations between 150 μM and 200 μM. As with any reagent used in PCR, dNTPs should be aliquoted and stored at $-20°C$.

REACTION BUFFER

The PCR reaction requires a complex buffer including the enzyme, dNTPs, primers, and Mg^{++} (cationic cofactor) in a Tris-buffer of appro-priate pH and salt concentration. Although for any given pair of oligo-nucleotide primers an optimal buffer condition can be established, there is no single set of conditions that will be optimal for all possible reactions. It is important to consider that, if appropriate conditions are not used, nonspecific products and primer-dimers will be formed. Both are substrates during the PCR reaction and will result in a lower yield of the desired product.

Taq polymerase requires Mg^{++} not bound to dNTPs or target DNA. A concentration of 0.2–2.5 mM Mg^{++} over the total dNTP concentration is recommended to guarantee its availability since dNTPs will bind free Mg^{++} at equimolar concentrations. Magnesium affects enzyme activity and fidelity, primer-template annealing, and template denaturation (Figure 3). The presence of EDTA or other chelators will lower the concentration of free Mg^{++} and will reduce the yield of the reaction [22]. The use of KCl and other salts, such as ammonium sulfate, also affect the denaturation and annealing temperature of DNA, as well as enzyme activity.

Some PCR protocols include dimethyl sulfoxide (DMSO) and glycerol to reduce the difficulties of polymerase extension due to secondary structure formation of the target DNA. The presence of 10% DMSO in PCR

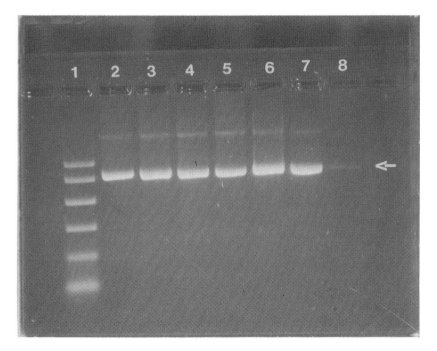

Figure 3 Effect of pH and Mg^{++} concentration on PCR specificity and yield. Molecular weight markers (lane 1); 1.5 mM, 2 mM, 2.5 mM, 3.5 mM Mg^{++} (pH 8.5) (lanes 2–5, respectively); pH 9.0, 9.5, 10.0 (Mg^{++} 2.0) (lanes 6–8, respectively). Amplification of a 600-bp fragment in a 100 µl final volume containing described buffer conditions, 200 µM each dNTP, 250 nM each primer, 5 ng target DNA, and 2.5 units Taq polymerase. Thirty cycles of amplification were performed and 10 µl of the reaction were loaded in a 2% agarose gel.

has been reported to enhance, or be essential for, the amplification of certain genes [34]. However, in some reactions the presence of DMSO rarely improved the quality or quantity of a product when that product was obtainable, even if poorly, without its addition. DMSO concentrations > 10% show considerable inhibition of Taq polymerase [31]. Formamide (1.25–5%) facilitates certain primer-template annealing and is also thought to lower the denaturing temperature of DNA. The presence of nonionic detergents (e.g., Tween 20, Nonidet P-40, Triton X-100) in the PCR reaction is known to reverse the inhibitory effects of ionic detergents such as SDS [35]. Also, they are important when working with environmental samples since their presence also can inhibit proteolytic enzymes that may persist in certain DNA preparations.

Other reagents, such as Perfect Match DNA® (Stratagene, CA), are polymerase enhancers that improve the specificity and yield of the PCR by destabilizing mismatched primer-template complexes and by helping to remove secondary structures that could impede normal extension [36]. Gelatin and bovine serum albumin (BSA) are enzyme stabilizers that also bind certain PCR inhibitors [21].

All buffer components are readily available from many commercial suppliers, individually or as a kit, and it is important that all reagents be of the highest quality available (molecular biology grade). Also available are commercial optimization kits (e.g., Stratagene, Invitrogen) that contain useful buffers and adjuncts specially selected for use in PCR and configured in a testing matrix that simplifies the optimization of reaction conditions. The following protocol is designed to determine the optimum concentration of Mg^{++} for a given set of primers and target DNA [37].

PROTOCOL II

(1) Prepare Mg-free master mix for eight reactions by mixing:
  - 80 μl 10X buffer [500 μM KCl, 1 mg/ml gelatin, 100 mM Tris-HCl (pH 8.4)]
  - 80 μl 2 mM dNTPs mix
  - 8 μl 50 μM primer 1
  - 8 μl 50 μM primer 2
  - 544 μl 15 μg/ml template genomic DNA, if plasmid, use less
(2) Prepare the following in seven microfuge tubes (0.5 ml):

Tube	H_2O (μl)	100 mM $MgCl_2$ (μl)
1	9.0	0.0
2	7.5	1.5
3	6.0	3.0

4	4.5	4.5
5	3.0	6.0
6	1.5	7.5
7	0.0	9.0

(3) To each tube in step 2 add 90 µl of mix from step 1. Add 1 µl of 2.5 U/µl Taq DNA polymerase to each tube.

(4) Mix, centrifuge briefly, and overlay with mineral oil.

(5) Denature the reaction at 96°C for two minutes.

(6) Perform thirty cycles of amplification using the following parameters:
- denaturation: 96°C, one minute
- annealing: 50°C, one minute
- extension: 72°C, two minutes

(7) Conclude with a final extension at 72°C for five minutes.

(8) An aliquot (10 µl) from each sample should be readily visible on an ethidium bromide-stained gel as a single band.

Examine gel to determine which $MgCl_2$ concentration results in the greatest amount of product. The protocol can be repeated using a narrower range of $MgCl_2$ concentrations to define an optimized concentration more precisely [37]. It is good practice to test the optimum magnesium concentration in a tube containing the environmental sample seeded with a known target as a control. This becomes very important when dealing with extremely inhibitory samples.

TARGET DNA

Target DNA sequences known sufficiently to design primers must be available to monitor microorganisms in environmental samples by PCR. The method has been successful in the detection of organisms from soil, water, and air samples. The major problem with the application of PCR to environmental samples is the presence of compounds inhibiting the reaction. It is critical to remove these inhibitors (i.e., humic acids, clay, organics) in order to have a successful amplification [38]. To enhance a successful amplification, it sometimes may be necessary to dilute the sample after the first few cycles of PCR, or after a completed reaction, and then perform an additional amplification. This may effectively dilute potential inhibitors to an acceptable level and allow for successful amplification. Various approaches are described in this book, depending on the type of sample and organisms.

Centrifugation or filtration is the most common method used for the recovery of cells from environmental samples. After these steps, cells can be lysed and nucleic acids purified [39], or cells can be eluted from the filters and used directly for PCR after a lysis step in the presence of, or without removing, the membrane [40]. Also, cells can be lysed on the membrane, and the nucleic acids fixed to the membrane and subsequently amplified [11]. With the latter protocol, termed "solid-phase PCR," several subsequent amplification reactions can be carried out on the same membrane without loss of specificity. This is unlike the liquid phase in which the templates can only be used once. For soil samples, bacterial cells are usually lysed in the samples and then the nucleic acids extracted [12]. The detection of microorganisms in aerosols has been accomplished by concentration of cells after impingement in liquid [3] and by direct concentration onto a membrane [41]. Standard purification protocols will probably not be effective for all environmental samples, and the required conditions must be determined on an individual basis. At least one copy of the target sequence should be intact after purification, over the length to be amplified, in order to proceed with the amplification.

Most PCR protocols require purified DNA that can be obtained from environmental samples by cell extraction followed by cell lysis and DNA purification or by cell rupture within the environmental matrix followed by DNA extraction and purification. Higher yields are obtained with the direct extraction methods, but the DNA may contain impurities and contamination that can inhibit the amplification. Samples prepared via standard molecular methodologies [42] are sufficiently pure for PCR, and usually no extra purification steps are required. Samples should be deproteinized (i.e., standard phenol-chloroform extraction) before introduction into the reaction mixture to guarantee removal of proteases and nucleases that may affect the reaction. However, successful amplification can be carried out in cell-free extracts [43].

STANDARD PROTOCOL FOR RNA AMPLIFICATION

The RT-PCR technique is a powerful method for the analysis of RNA transcripts. A cDNA copy of the RNA may be synthesized using random hexamers and reverse transcriptase. Other options for priming include oligo (dT) or specific downstream primers. It has been shown that the random-hexamer approach is the most consistent and results in the highest amplification of target sequences [44]. Residual DNA may be digested with RNase-free DNase prior to reverse transcription so that it does not contribute to the final PCR product, and appropriate controls (i.e., minus RT reactions) should be used in each experiment. The single-stranded cDNA product can then be amplified by the PCR after heat inactivation of the reverse transcriptase and adjustment of the reaction to PCR conditions.

As with DNA amplification, it is recommended that each component of the reaction be optimized.

Amplification from mRNA rather than DNA allows the analysis of gene expression and may be used to correlate viability of the target organism. The application of this technique includes the amplification of cDNA derived from all types of RNA: mRNA, rRNA, total RNA, and RNA viruses. Traditionally, mRNA or cellular RNA has been detected or monitored by methods such as Northern and Southern blotting or radioimmunoassays. Currently, PCR is the practical tool for these studies due to orders of magnitude of greater sensitivity. Methods for RNA extraction are described in detail in several laboratory manuals [37,42], and most of them are PCR compatible. The only step different from a typical PCR reaction is the required conversion of the RNA template to cDNA with reverse transcription (Figure 4). The same PCR buffer can be used for the RT and amplification steps.

The RT-PCR can now be done in a one-step, single-enzyme, single-tube reaction, with the use of a heat-stable reverse transcriptase. A new thermostable enzyme purified from *Thermus thermophilus* (Tth) is able to efficiently reverse transcribe RNA to cDNA in the presence of $MnCl_2$ at elevated temperatures, and to act as a DNA polymerase for subsequent PCR amplification in the presence of $MgCl_2$, after chelation of the manganese ion with EGTA (Perkin-Elmer, CT). This amplification requires neither the addition of a second enzyme nor a modification of the buffering conditions for transition from reverse transcription to PCR.

Reagents and solutions used for RT-PCR procedures should be prepared using standard methods for handling RNA [42]. All solutions should be treated with diethylpyrocarbonate (DEPC) to get rid of RNAse contamination. The DEPC reacts with histidine residues of proteins and will inactivate RNAses. However, it can also react with RNA. Therefore, DEPC needs to be properly applied and removed before the solution is used. It is important to be aware of the risk of RNAse contamination, both during work and storage of RNA samples. Gloves must be worn when handling any reagent or reaction vessel. In addition, the use of RNAsin or a similar RNAse inhibitor is necessary when dealing with environmental samples. The following is a general protocol for RNA amplification.

PROTOCOL III

(1) Set up a 21-μl reaction in a 0.5-ml microfuge tube containing:
- 5 μl RNA (0.5–1 μg)
- 5 μl random hexamers (110 pmol)
- 11 μl dH$_2$O
(2) Heat the reaction ten minutes at 70°C and quick-chill on ice.

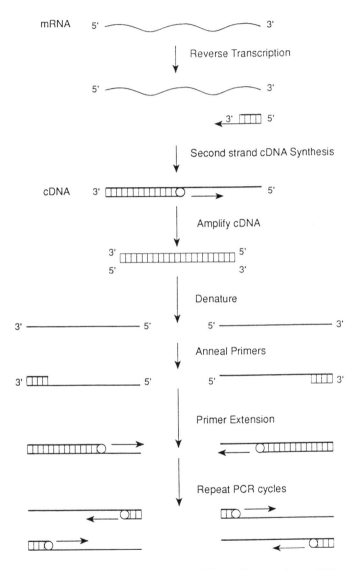

Figure 4 Diagram of the main steps of a PCR amplification of target RNA.

(3) Add to the reaction:
- 5 μl 10X PCR buffer
- 4 μl dNTP mixture (1 mM each dNTPs)
- 1 μl RNAsin (Promega), 1 U/μl
- 1 μl Mo-MuLV reverse transcriptase, 100–200 U
- 18 μl dH$_2$O

(4) Mix and incubate ten minutes at room temp., then thirty to sixty minutes at 42°C.

(5) To inactivate the reverse transcriptase, heat the reaction at 95°C for five to ten minutes and quick-chill on ice.

(6) Set up a standard PCR reaction using as template 5–50 μl of this mixture. The optimal amount should be determined in each case.

PRODUCT (AMPLICON) DETECTION

To demonstrate the presence of amplified products, the detection formats frequently used are gel electrophoresis along with hybridization. Since the reliability of PCR and the amount of amplified product are highly satisfactory, in most cases all that is necessary is a simple gel electrophoresis. Purification of the products before detection is optional and can be done with a standard phenol-chloroform extraction followed by ethanol precipitation [42] or using prepacked commercial PCR purification columns. Only a small aliquot of the amplified product is needed for an electrophoresis, usually 1/10 of the reaction, but care should be taken to avoid the mineral oil that can interfere with the procedure. Usually the solution is obtained through the oil layer and the tip wiped. Similarly, the sample can be frozen and the oil removed.

In gel electrophoresis, the defined length of the amplification product should agree with the expected size according to the annealing sites of the primers when the gel is observed under UV light after ethidium-bromide staining. Primer-dimers, and sometimes the primers themselves (in case of an overload) may be seen in the gel, but most products should be a sharp, single band. Multiplex amplification products are an exception, since multiple products are expected. Nonspecific amplification products, due to primers annealing to nonspecific sites, lower or higher primer concentrations, or unequal molar amounts of primers, may be seen in the gel. Optimization of the amplification conditions should be performed if this is observed. Generally, PCR products of 150–800 bp are produced with the advantage of shortening the time of amplification, since products in this size range can be produced using a two-step amplification cycling. Polyacrylamide gels give greater resolution than agarose gels for PCR products and small DNA fragments, however, the preparation of these gels can be tedious and time-consuming. Agarose formulations specially designed for separation of small DNA fragments are now available that not only offer speed and convenience but also a resolution almost equal to that of polyacrylamide. The detection of amplified products based only on the expected length in an electrophoresis does not always provide unequivocal positive identification. Products of the reaction may be sequenced, treated with restriction enzymes, or hybridized to specific gene probes.

Probe hybridization to an internal product sequence between those of the two primers, following a Southern, dot, or reverse dot blotting step [14], improves not only the specificity of the reaction, but also the sensitivity. This, combined with the advantages of using more sample volume and a simpler, faster process, makes dot blotting a common choice for detection. However, one should realize that dot blotting alone is not, in some cases, definite proof of a successful amplification due to the fact that no molecular weight reference is available when only this type of blotting is used. A second possibility involves recovering the specific fragment from the gel (e.g., electroelution) and then blotting. For these protocols, the transfer and hybridization conditions should be optimized for short DNA fragments. It must be kept in mind that nucleic acid transfer by Southern blotting is not 100% efficient [45].

Strong signals are usually obtained using isotopic-labeled probes; however, PCR produces products in sufficient quantity to allow the use of nonisotopical detection methods. Nonisotopic detection methods have many advantages over isotopic methods and are very practical [46]. In addition to comparable sensitivity and a decrease in the amount of the time required for detection, there are no problems associated with the handling and disposal of radioactive material. A wide variety of different nonradioactive labels are available, including biotin, digoxigenin, fluorescent dyes, and chemiluminescence kits from various commercial suppliers. Specific, internal gene probes can be labeled, both isotopically or nonisotopically, but PCR products can also be labeled during the amplification process. Labeled dNTPs or primers, via $5'$ end-labeling with T4 polynucleotide kinase, can be incorporated to almost any desired specific activity to the basic reaction. In this case, the sensitivity of the reaction will be directly increased after the appropriate detection method.

Other detection methods, such as HPLC [47] and capillary electrophoresis [48], offer resolution approaching, and sensitivity exceeding, gel electrophoresis. Although expensive, advantages of these methods include the opportunities for speed, automation, and quantification, which are not obtained with conventional approaches.

CONTAMINATION CONTROL

One of the greatest problems facing PCR is false positivity due to contaminating target DNA from previous amplifications. The problem is especially acute in optimized reactions. Due to the high sensitivity of PCR, one molecule of DNA is enough to contaminate the reaction. Sources of contamination include product carryover (accumulation of PCR products in the laboratory after repeated amplifications of the same target sequence) and exogenous DNA (stock solution of plasmid or genomic DNA). Appropriate pre- and postamplification precautions need to be taken to prevent contamination problems. Sometimes the amount of contamination in a sample is so small that it cannot be seen by gel electrophoresis and will be detected only after hybridization of the products.

In general, avoiding human errors, planning the experiments in advance, and good laboratory procedures will reduce the risk of contamination. All solutions should be prepared free from contaminating nucleic acids and/or nucleases (DNAses/RNAses) using the highest quality components, including water. Water can be a factor affecting reproducibility and yield of the reaction and should be of the highest quality, distilled/deionized, filtered and autoclaved. Use of gloves, disposable sterile bottles or tubes, plugged pipette tips, and autoclavable positive-displacement pipettes should be considered common practice. Aliquoting reagents into single- or two-time usage volumes for storage can be a factor affecting consistency from experiment to experiment. An area should be set aside specifically for manipulation of sample postamplification.

Aerosolization of PCR products can occur each time the tubes are opened and pipetted; therefore, samples should be spun before opening and handled carefully. Working in laminar flow-hood or workstation "still air enclosure" acrylic chambers (e.g., Template-Tamer, Coy) avoids spread of aerosolized particles in air currents and they are easy to decontaminate. Saksena et al. [49], assessed the utility of a "sentinel" tube technique (an empty, uncapped, sterilized microfuge tube left open) in evaluating and monitoring laboratory contamination by aerosols.

If contamination is suspected, control measures have to be taken. For example, discard rather than attempt to decontaminate suspected aliquots, autoclave tubes and tips, and depurinate any residual DNA (0.1 M HCl solution) on surfaces or equipment. Another approach to PCR product decontamination involves short wavelength (254 nm) ultraviolet irradiation of the reaction mixture, minus target and thermostable polymerase, prior to amplification [50,51]. A 10% working solution of bleach can also be used to treat surfaces to prevent carryover. However, a rinse using 1% sterile sodium thiosulfate should be used to eliminate any residual chlorine.

Two methods can be used to prevent amplification of carryover products in the laboratory. The first one is enzymatic, which does not prevent accumulation of the product in the laboratory, but avoids reamplification and, thus, false positives [52]. Deoxyuridine triphosphate (dUTP) is substituted for deoxythymidine triphosphate (dTTP) in all PCR reactions. The enzyme uracil *N*-glycosylase (UNG) catalyzes the excision of uracil from any potential single- or double-stranded PCR carryover DNA present in the reaction prior to the first PCR cycle and, thus, will render them ineligible to serve as templates for subsequent amplifications. Appropriate steps should be taken to avoid any degradation of new products due to any residual activity of UNG. Incomplete inactivation of UNG at high temperatures and lower amplification efficiency due to the change of dTTP for dUTP are the most common difficulties associated with this method.

The second option, photochemical modification of amplified DNA [53], prevents any subsequent amplification of the products. After the incorporation of isopsoralen reagents into the amplification, the products are exposed to longwave UV light, which photochemically activates the isopsoralen, chemically modifying the products and thereby blocking the polymerase from further extension after it encounters a modified base in the template strand. Inhibition of amplification has been observed on occasion; however, isopsoralens can be included in PCR reactions under conditions chosen to minimize their effect on amplification.

POSITIVE AND NEGATIVE CONTROLS

When analyzing environmental samples, it becomes extremely important to include positive and negative controls as an intrinsic part of the procedure to ensure the validity of the experiment. As a result of the complex nature of most environmental samples, it is very difficult, if not impossible, to completely remove any possible inhibitory substances. It should be realized that each inhibitory compound will act differently on the enzyme or other components of the reaction mix. The enzyme may be completely poisoned, or the V_{max} may be changed as a result of the inhibitory substances. Although it would be unnecessary and impractical to carry out kinetic studies on each individual sample to determine the best reaction conditions, the possible inhibitory effects of the background chemistry of the sample needs to be addressed.

The negative as well as the positive controls need to be carefully chosen. If any of the controls is contaminated, the experiment should be discarded and the source of contamination detected and eliminated. For positive controls, the amount of target being used should be minimized for two reasons. First, to serve as a control, a limited number of copies of the target should be amplified weakly but consistently in an optimized protocol. Second, by using a high dilution of DNA, there is less chance of contamination of other samples after amplification. Although the use of a purified target DNA in a separate tube can be a useful control, it only gives information on the activity of the enzyme and the status of the PCR mix components. In most cases this is ineffective in assessing the true inhibitory effects of environmental samples. Thus, it becomes necessary to design other possible controls, such as the possibilities presented below:

(1) A standardized low concentration of purified target DNA could be added to a PCR reaction tube containing an aliquot of the sample. This tube is then subjected to amplification parallel to the sample. If the control DNA fails to be amplified, this could only mean that the amplification reaction is being inhibited. Thus, the sample should be subjected to further purification procedures prior to amplification.

(2) An internal control could be included in each of the test amplifications. The internal control could be a segment of the target DNA, synthesized taking into consideration the sequence and the molecular weight of the target DNA. A possible construct might consist of the target sequence with a deleted internal segment (Colin Fricker, personal communication). If this construct is added at the lowest possible concentration, then the primers should use it as a template for the amplification of a segment that is shorter than the experimental target sequence. In this manner, if both the control and the target are amplified, they can both be visualized as discrete and separate bands.

Although both types of positive controls are useful, they each have potential pitfalls. The use of a separate tube entails the use of more materials and the possibility of missing the detection of the target DNA in cases when it is present in extremely low concentrations. Since the molecular weight of the control amplicon will be the same as that of the target in the sample, and, if the target is present in only a few copies, there might not be any apparent amplification in the sample tube. The use of an internal control with a deleted segment might under some circumstances lead to false negatives. This is a result of the more efficient PCR amplification of shorter-over-longer segments of DNA. A possible method to circumvent this problem might be to use an internal control with an internal unique sequence ligated to the target DNA, which results in a longer amplicon product. This could be a more stringent control as a result of the longer segment to be amplified.

(3) A more accessible control to any environmental scientist might be the inclusion of an internal control that corresponds to a sequence present in all eubacteria. The primers (5′-TCAgAWYgAACgCTggCgg-3′ and 5′-AAggAggTgATCCAgCC-3′) that allow for amplification of a universally conserved eubacterial 16S rRNA should be the perfect internal control for any environmental sample. These primers yield a 1530 bp amplicon [54] and could, therefore, easily be separated from the target DNA by electrophoresis. Additionally, if the size of the

target sequence is approximately that of the internal control, restriction analysis (such as ARDRA) of the products will give unequivocal results. A negative control will allow for the determination of contamination of the sample with target sequences; however, negative controls are not meaningful unless there is also a positive control being run in parallel. A typical negative control might be a parallel amplification tube without target DNA or containing an autoclaved portion of the sample plus the reaction mix. If there is amplification, this could only mean extraneous contamination.

SUMMARY

The PCR assay provides a unique opportunity to indirectly detect extremely low concentrations of microorganisms in environmental samples. The optimal conditions under which PCR amplification works best are known, and it seems reasonable to assume that no such standard conditions will be established for all possible reactions. After all, researchers use PCR for very different purposes, and each environmental sample is unique. There are some things that are generally true that may be useful to keep in mind when setting up a particular reaction. This chapter presented the basis of the PCR reaction and a strategy for a start-up amplification of a given target sequence. The general conditions described in this chapter have usually provided satisfactory results. However, minor adjustments to these conditions are likely to improve a marginal PCR reaction into one with excellent specificity and enhanced sensitivity of detection.

REFERENCES

1 Ward, R. L. and E. W. Akin. 1984. Minimum infective dose of animal viruses. *CRC Crit. Rev. Environ. Control.* 14:297–310.
2 Persing, D. H. 1991. Polymerase chain reaction: Trenches to benches. *J. Clin. Microbiol.* 29:1281–1285.
3 Alvarez, A. J., M. P. Buttner, G. A. Toranzos, E. A. Dvorsky, A. Toro, T. B. Heikes, L. E. Mertikas-Pifer, and L. D. Stetzenbach. 1994. Use of solid-phase PCR for enhanced detection of airborne microorganisms. *Appl. Environ. Microbiol.* 60:374–376.
4 Atlas, R. M. and A. K. Bej. 1990. Detecting bacterial pathogens in environmental water samples by using PCR and gene probes, pp. 399–407. In: M. Innis, D. Gelfand, D. Sninsky, and T. White (eds.), *PCR protocols: A guide to methods and application.* Academic Press, New York.
5 Bej, A. K., J. L. DiCesare, L. Haff, and R. M. Atlas. 1991. Detection of

Escherichia coli and *Shigella* spp. in water by polymerase chain reaction and gene probes for *uid*. *Appl. Environ. Microbiol.* 57:1013–1017.

6 Chaudhry, G. R., G. A. Toranzos, and A. R. Bhatti. 1989. Novel method for monitoring genetically engineered microorganisms in the environment. *Appl. Environ. Microbiol.* 55:1301–1304.

7 Mahbubani, M. H., A. K. Bej, R. Miller, L. Haff, and R. M. Atlas. 1990. Detection of bacterial mRNA using PCR. *BioTechniques.* 10:48–49.

8 Pollard, P. R., W. M. Johnson, H. Lior, S. D. Tyler, and K. R. Rozee. 1990. Detection of the aerolysin gene in *Aeromonas hydrophila* by the polymerase chain reaction. *J. Clin. Microbiol.* 28:2477–2481.

9 Starnbach, M. N., S. Falkow, and L. S. Tompkins. 1989. Species specific detection of *Legionella pneumophila* in water by DNA amplification and hybridization. *J. Clin. Microbiol.* 27:1257–1261.

10 Steffan, R. J. and R. M. Atlas. 1988. DNA amplification to enhance the detection of genetically engineered bacteria in environmental samples. *Appl. Environ. Microbiol.* 54:2185–2191.

11 Toranzos, G. A., and A. J. Alvarez. 1992. Solid-phase PCR for the detection of pathogens in water. *Can. J. Microbiol.* 38:365–369.

12 Tsai, Y. L. and B. H. Olson. 1992. Detection of low numbers of bacterial cells in soils and sediments by polymerase chain reaction. *Appl. Environ. Microbiol.* 58:754–757.

13 Mullis, K., F. Faloona, S. Scharf, R. Saiki, G. Horn, and H. Erlich. 1986. Specific enzymatic amplification of DNA in vitro: The polymerase chain reaction. *Cold Spring Harbor Symp. Quant. Biol.* 51:263–273.

14 Saiki, R. K., D. H. Glenfand, S. Stoffel, S. J. Scharf, R. Higuchi, G. T. Horn, K. B. Mullis, and H. A. Erlich. 1988. Primer-directed enzymatic amplification of DNA with a thermostable DNA polymerase. *Science.* 239:487–494.

15 Ochman, H., J. W. Ajioka, D. Garza, and D. L. Hartl. 1989. Inverse polymerase chain reaction. In H. A. Erlich (ed.), *PCR Technology.* Stockton Press, New York. pp. 105–111.

16 Nelson, R. M. and G. L. Long. 1989. A general method of site-specific mutagenesis using a modification of the *Thermus aquaticus* polymerase chain reaction. *Anal. Biochem.* 180:147–151.

17 Wang, A. M. and M. V. Mark. 1990. Quantitative PCR. pp. 70–75. In Innis, M.A., D.H. Gelfand, J.J. Sninsky, and T.J. White (eds.). *PCR protocols: A guide to methods and applications.* Academic Press, Inc. San Diego, Calif.

18 Stofle, E. S., D. D. Koeberl, G. Sakar, and S. S. Sommer. 1988. Genomic amplification with transcript sequencing. *Science.* 239:491–494.

19 Bej, A. K., S. C. McCarty, and R. M. Atlas. 1991. Detection of coliform bacteria and *Escherichia coli* by multiplex polymerase chain reaction: Comparison with defined substrate and plating methods for water quality monitoring. *Appl. Environ. Microbiol.* 57:2429–2432.

20 Chamberlain, J. S., R. A. Gibbs, J. E. Ranier, P. N. Nguyen, and C. T. Cashey. 1988. Deletion screening of the Duchenne muscular dystrophy locus via multiplex DNA amplification. *Nucleic Acids Res.* 16:11141–11156.

21 Cha, R. S. and W. G. Thilly. 1993. Specificity, efficiency, and fidelity of PCR. *PCR Methods and Applications.* 3:S18–S29.

22 Saiki, R. K. 1992. The design and optimization of the polymerase chain reaction. In H. A. Erlich (ed.), *PCR technology.* W. H. Freeman, NY. pp. 7–16.

23 Ruano, G., W. Fenton, and K. K. Kidd. 1989. Biphasic amplification of very dilute DNA samples via "booster" PCR. *Nucleic Acids Res.* 17:5407.

24 Pierre, C., D. Lecosier, Y. Boussougant, D. Bocart, V. Joly, P. Yeni, and A. J. Hance. 1991. Use of a reamplification protocol improves sensitivity of detection of *Mycobacterium tuberculosis* in clinical samples by amplification of DNA. *J. Clin. Microbiol.* 29:712–717.

25 Mullis, K. and F. Faloona. 1987. Specific synthesis of DNA in vitro via a polymerase-catalysed chain reaction. *Methods Enzymol.* 155:335–350.

26 Albert, J. and E. M. Fenyo. 1990. Simple, sensitive and specific detection of human immunodeficiency virus type 1 in clinical specimens by polymerase chain reaction with nested primers. *J. Clin. Microbiol.* 28:1560–1564.

27 Erlich, H. A., F. Gelfand, and J. J. Sninsky. 1991. Recent advances in the polymerase chain reaction. *Science.* 252:1643–1651.

28 Schwarz, K., T. Hansen-Hagge, and C. Bartram. 1990. Improved yields of long PCR products using gene 32 protein. *Nucleic Acids Res.* 18:1079–1082.

29 Linz, U. 1990. Thermocycler temperature variation invalidates PCR results. *Biotechniques.* 9:286–293.

30 Saiki, R. K., S. Scharf, F. Faloona, K. B. Mullis, G. T. Horn, H. A. Erlich, and N. Arnheim. 1985. Enzymatic amplification of β-globin genomic sequences and restriction site analysis for diagnosis of sickle cell anemia. *Science.* 230:1350–1354.

31 Gelfand, D. H. 1992. Taq DNA polymerase. In H.A. Erlich (ed.), *PCR technology.* W.H. Freeman, NY. pp. 17–22.

32 Tindall, K. R. and T. A. Kunkel. 1988. Fidelity of DNA synthesis by the *Thermus aquaticus* DNA polymerase. *Biochemistry,* 27:6008–6013.

33 Dieffenbach, C. W., T. M. J. Lowe, and G. S. Dveksler. 1993. General concepts for PCR primer design. *PCR Methods and Applications.* 3:S30–S37.

34 Filichkin, S. A. and S. B. Gelvin. 1992. Effect of dimethyl sulfoxide concentration on specificity of primer matching in PCR. *Biotechniques.* 12:828–830.

35 Weyant, R. S., P. Edmonds, and B. Swaminathan. 1990. Effect of ionic and nonionic detergents on the Taq polymerase. *BioTechniques.* 9:308–309.

36 Nielson, K. and E. J. Mathur. 1990. Perfect match enhancer: Limits false priming events during amplification reactions. *Strategies in Molecular Biology.* 3:17–22.

37 Ausubel, F. M., R. Brent, R. E. Kingston, D. D. Moore, J. G. Seidman, J. A. Smith, and K. Struhl. 1992. *Short protocols in molecular biology.* John Wiley & Sons, New York.

38 Tsai, Y. L. and B. H. Olson. 1992. Rapid method for separation of bacterial DNA from humic substances in sediments for polymerase chain reaction. *Appl. Environ. Microbiol.* 58:2292–2295.

39 Somersville, C. C., I. T. Knight, W. L. Straube, and R. R. Colwell. 1989. Simple, rapid method for direct isolation of nucleic acids from aquatic environments. *Appl. Environ. Microbiol.* 55:548–554.

40 Bej, A. K., M. H. Mahbubani, J. L. DiCesare, and R. M. Atlas. 1991. PCR-gene probe detection of microorganisms using filter concentrated samples. *Appl. Environ. Microbiol.* 57:3529–3534.

41 Sawyer, M. H., C. J. Chamberlin, Y. N. Wu, N. Aintablian, and M. R. Wallace. 1994. Detection of varicella-zoster virus DNA in air samples from hospital rooms. *J. Infect. Dis.* 169:91–94.

42 Sambrook, J., E. F. Fritsch, and T. Maniatis. 1989. *Molecular cloning: A laboratory manual.* Cold Spring Harbor Laboratory Press, Cold Spring Harbor, New York.

43 deBruijn, F. J. 1992. Use of repetitive (repetitive extragenic palindromic and enterobacterial repetitive intergeneric consensus) sequences and the polymerase chain reaction to fingerprint the genomes of *Rhizobium meliloti* isolates and other soil bacteria. *Appl. Environ. Microbiol.* 58:2180–2187.

44 Noonan, K. E. and I. B. Roninson. 1988. mRNA Phenotyping by enzymatic amplification of randomly primed cDNA. *Nucleic Acids Res.* 16:10366–10368.

45 Selder, R. 1985. DNA probes for microbial diagnosis. *Med. Lab. Sci.* 42:352–360.

46 Kricka, L. J. 1992. Nucleic acid hybridization test formats: strategies and applications. In: L.J. Kricka (ed.), *Nonisotopic DNA probe techniques.* Academic Press, NY. pp. 3–28.

47 Katz, E. D. and M. W. Dong. 1990. Rapid analysis and purification of polymerase chain reaction products by high-performance liquid chromatography. *Biotechniques.* 8:546–555.

48 Guttman, A. 1994. Separation of DNA by capillary electrophoresis. In J. P. Landers (ed.), *Handbook of capillary electrophoresis.* CRC Press, Boca Raton, FL. pp. 129–143.

49 Saksena, N. K., D. Dwyer, and F. Barre-Sinoussi. 1991. A "sentinel" technique for monitoring viral aerosol contamination. *J. Infect. Dis.* 164:1021–1022.

50 Ou, C. Y., J. L. Moore, and G. Schochetman. 1991. Use of UV irradiation to reduce false positivity in polymerase chain reaction. *BioTechniques,* 10:442–445.

51 Sarkar, G. and S. S. Sommer. 1990. More light on PCR contamination. *Nature (London).* 347:340–341.

52 Longo, M. C., M. S. Berninger, and J. L. Hartley. 1990. Use of uracil DNA glycosylase to control carry-over contamination in polymerase chain reaction. *Gene.* 93:125–128.

53 Jinno, Y., K. Yoshiiura, and N. Niikawa. 1990. Use of psoralen as extinguisher of contaminated DNA in PCR. *Nucleic Acids Res.* 18:6739.

54 Liesack, W., H. Wyland, and E. Stackebrandt. 1991. Potential risks of gene amplification by PCR as determined by 16S rDNA analysis of a mixed-culture of strict barophilic bacteria. *Microb. Ecol.* 21:191–198.

CHAPTER 3

Assessing Genetic Diversity of Microbes Using Repetitive Sequence-Based PCR (rep-PCR)

FRANK J. LOUWS[1,2]
MARIA SCHNEIDER[1,2]
FRANS J. DE BRUIJN[1,2,3]

INTRODUCTION

MICROBES are becoming increasingly important in ecological, industrial, and agricultural applications. Specific bacterial strains have been selected and manipulated for bioremediation, such as reducing toxicants in landfill waste sites, and for novel industrial applications, such as harvesting of enzymes, manufacturing plastics, and biomass conversion for fuel or compost. In addition, microbes are increasingly employed for agricultural applications, including plant growth promotion, enhanced nitrogen fixation and biological control of pests. Microbes, as contaminants or pathogens, are also economically important in a negative fashion. For example, phytobacteria limit optimum productivity of crops useful for food, fiber, and alternative energy sources.

The increased exploitation of beneficial microbes and the need to control bacterial pathogens has been accompanied by the need for biological and technological advances in the area of bacterial strain identification (diagnosis) and classification. In collaboration with the group led by Dr. J. R. Lupski at Baylor College of Medicine (Houston, TX), DNA-based protocols have been developed to fingerprint the genomes of many Gram-negative and Gram-positive bacteria of medical, ecological, industrial, and agricultural importance [1–10].

[1]National Science Foundation Center for Microbial Ecology, Michigan State University, East Lansing, MI 48824, U.S.A.
[2]MSU-DOE Plant Research Laboratory, East Lansing, MI 48824, U.S.A.
[3]Department of Microbiology, Michigan State University, East Lansing, MI 48824, U.S.A.

The technique employs primers that correspond to *rep*etitive DNA sequences, and, through *PCR* (rep-PCR), numerous DNA bands of variable length are selectively amplified. Amplified bands can be size fractionated through a gel matrix to yield complex fingerprint patterns. Rep-PCR has been used to assess the genetic diversity of microbial communities, to classify or differentiate closely related strains, to diagnose plant pathogenic and symbiotic microbes, and to conduct environmental or epidemiological studies (see Reference [11]).

Also, rep-PCR has proven useful for classifying, identifying, and differentiating medically important microbes [1,3,6–8,11–15]. The purpose of this chapter is to outline the general utility of rep-PCR and to provide details to facilitate the successful adaption of rep-PCR to experimental problems and challenges in other laboratories.

TOOLS FOR IDENTIFYING AND CHARACTERIZING BACTERIA

Historically, a variety of phenotypic, biochemical, and serological characteristics have been employed to classify and identify bacteria. Techniques, such as serology-based methods (ELISA), substrate utilization screens (BIOLOG), fatty-acid analyses (FAME), and multilocus enzyme electrophoresis (MLEE), have been reliably exploited for strain identification. However, phenotypic features often lack stability and, perhaps with the exception of MLEE, may not necessarily reflect the genetic relatedness of strains. In contrast, DNA-based protocols appear to be more stable, more closely reflect phylogenetic relationships, and are useful for arranging strains into coherent groups. In fact, DNA-DNA hybridization protocols form the basis for delimiting geno-species of bacteria [16], but they lack the sensitivity to differentiate closely related strains.

This sensitivity can be enhanced by employing other DNA-based techniques, including digestion of total DNA using endonucleases that frequently (restriction enzyme analysis, REA) or infrequently (pulsed-field gel electrophoresis, PFGE; field-inversion gel electrophoresis, FIGE) cut a genome, plasmid profiling, restriction fragment length polymorphism (RFLP) analyses, and 16S ribosomal gene analyses (e.g., rrn-RFLP, sequencing). The relative utility of these protocols for differentiating a genus, species, or strain is highlighted in Figure 1.

Several of the more recently developed methods for classification and strain identification outlined in Figure 1 are based on the amplification of DNA using the polymerase chain reaction. In general, two broadly defined classes of PCR-based protocols are used. The first class involves the amplification of one or a few specific DNA fragments for classification or identification. One example is the amplification of portions of the rRNA operon (for review, see Reference [17]) usually followed by DNA sequence

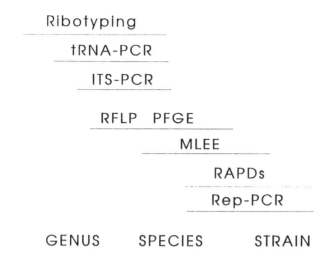

Ribotyping

tRNA-PCR

ITS-PCR

RFLP   PFGE

MLEE

Rapds

Rep-PCR

GENUS   SPECIES   STRAIN

Figure 1 Relative effectiveness of various protocols to assess diversity among bacteria at the genus, species, or strain level. Ribotyping includes the PCR amplification of portions of the ribosomal genes (usually 16S) followed by DNA sequence analysis or restriction digestion (ARDRA) [17,18]. The tRNA-PCR employs conserved primers to amplify t-RNA regions, and length polymorphisms are genus and species specific [20]. The ITS-PCR exploits species-specific length polymorphisms of the intervening DNA between the 16S and 23S RNA genes [19]. Restriction fragment length polymorphisms (RFLP) employs labeled probes of known or unknown sequence that hybridize to specific bands of DNA bound to a membrane. Total DNA is digested with a restriction enzyme and resolved on a matrix, then transferred to the membrane. The utility of RFLP for differentiating species and strains is dependent on the probe used. Large DNA fragments digested from total DNA with infrequently cutting restriction enzymes can be resolved by pulsed field gel electrophoresis (PFGE) and provide strain-specific and species-specific fingerprint profiles.

Multilocus enzyme electrophoresis employs starch gels to resolve strain-specific polymorphisms (isozymes) in migration rates of functioning enzymes. Assays employing many enzymes can be used effectively to assess the population structure of microbial communities [55]. Amplification of random DNA fragments using primers of arbitrary sequence (RAPDs) [21,22], or primers that correspond to repetitive sequences (rep-PCR, this chapter) are particularly useful for the analysis of closely related strains but may have some utility at the species level as well, depending on the structure of the population.

or restriction analysis (ARDRA) [18]. Second, ITS-PCR exploits the variable length of DNA segments that span the intergenic region between the 16S and 23S RNA genes [19]. Third, tRNA analysis has been employed for genus- and species-specific amplification of variable-length PCR products, employing primers that amplify specific tRNA regions [20]. Fourth, knowledge of the nature of specific DNA sequences has enabled the amplification of recognizable DNA segments that are diagnostic for a genus, species, or strain.

The second broad class of PCR-based protocols includes the use of short

DNA sequences that serve as primers to amplify numerous products from genomic DNA. When electrophoretically size fractionated, the resulting DNA fragments yield species- or strain-specific "patterns." In the case of microbes, two general protocols have been developed. The first protocol employs short primers of arbitrary sequence (RAPDs; [21]) or longer primers combined with low annealing temperatures (AP-PCR; [22]) to amplify anonymous DNA fragments of variable lengths. The RAPD or AP-PCR analysis has been adopted for the analysis of many different species and requires no specific prior knowledge of the nature of DNA sequences in the genome of the microbe to be studied. In addition, primers corresponding to the M13 core sequence have been exploited for arbitrary amplification of rather complex fingerprint patterns from DNA of numerous organisms, including eukaryotic microbes (fungi). Recent reviews highlight the utility of such protocols for microbes [23–25].

The second protocol employs primers that correspond to specific polynucleotide DNA sequences or repetitive DNA sequences. For example, PCR amplification using the polynucleotide primer GTGGTGGTG-GTGGTG generates complex fingerprint patterns from genomic DNA of fungi [25] and bacteria [11].

Alternatively, naturally occurring repetitive DNA elements can serve as primer binding sites for genomic DNA amplification [1,11]. The latter approach is similar to that used to fingerprint mammalian genomes, such as human *Alu*PCR (for a recent review, see Reference [26]). The focus of this chapter is on the use of repetitive DNA sequences for genomic fingerprinting.

BASIS FOR REPETITIVE SEQUENCE-BASED PCR (rep-PCR)

Several families of repetitive sequences are interspersed throughout the genome of diverse bacterial species (for recent reviews, see References [11,12]. The precise function of such sequences is not known [12], but they have been postulated to be important for chromosomal organization [27,28], gene expression, mRNA stability or protein-DNA interactions [29–33]. Regardless of their function, repetitive sequences appear to be located at "characteristic" positions in microbial genomes [34] and have turned out to be very useful to characterize genomic structure via genomic fingerprinting. Several families of repetitive sequences have been exploited to generate fingerprint profiles using PCR [11]. Versalovic et al. (1994) have detailed the design of rep-PCR and the different types of primers that can be used. This chapter highlights the most commonly used protocol and emphasizes the use of REP, ERIC, and BOXA1R primers (Table 1) for fingerprinting bacteria of environmental and especially agricultural importance.

TABLE 1. Oligonucleotide Primer Sequences and Annealing
Temperatures for REP-, ERIC-, and BOX-PCR.

Protocol	Primer	Sequence	Annealing Temperature (°C)
Rep-PCR[a]	REP1R	5′-IIIICgICgICATCIggC-3′	40
	REP2I	5′-ICgICTTATCIggCCTAC-3′	40
ERIC-PCR[a]	ERIC1R	5′-ATgTAAgCTCCTggggATTCACp3′	52
	ERIC2	5′-AAgTAAgTgACTggggTgAgCg-3′	52
BOX-PCR[b]	BOXA1R	5′-CTACggCAAggCgACgCTgACg-3′	53

[a]Versalovic et al., 1991 [1].
[b]Versalovic et al., 1994 [11].

The three families of repetitive sequences studied in most detail, including the 35–40 bp repetitive extragenic palindromic (REP) sequence [1,35,36], the 124–127 bp enterobacterial repetitive intergenic consensus (ERIC) sequence [1,37,38], and the 154 bp BOX element, comprised of three subunits (boxA, boxB and boxC; [15,39]) are unrelated at their DNA sequence level. The REP, as well as a centrally conserved sequence within ERIC and the boxA subunit, have inverted repeats, with a potential to form stable, stem-loop structures. The repetitive elements may be present in both orientations on the chromosome, and the conserved regions can serve as primer binding sites. The PCR primers have been designed to "read outward" from the inverted repeats, leading to the selective amplification of distinct genomic regions located between REP, ERIC, or BOX sequences (see Figure 2). The corresponding protocols are referred to as REP-PCR, ERIC-PCR, and BOX-PCR, respectively, and rep-PCR generally. The resolved patterns resemble a "bar code," analogous to UPC codes used in grocery stores, and function as a signature or fingerprint for specific bacterial strains (see Figure 3).

METHODS

The rep-PCR method is rapid and can be used routinely. Eighty to 100 samples can be prepared in fewer than two hours using standard pipetters and microfuge tubes. Efficiency can be enhanced by automation of the upstream reaction setup [40], including the use of 96-well plates and a robotic pipetting station. The samples are put in a thermocycler and subjected to 30–35 PCR cycles, requiring six to seven hours. For standard agarose gel electrophoresis, twenty minutes is required to load a 30-sample gel and electrophoresis requires six to eighteen hours, depending on the fragment resolution and discrimination desired. Therefore, a rep-

AMPLIFICATION OF DNA BETWEEN ADJACENT REPETITIVE ELEMENTS

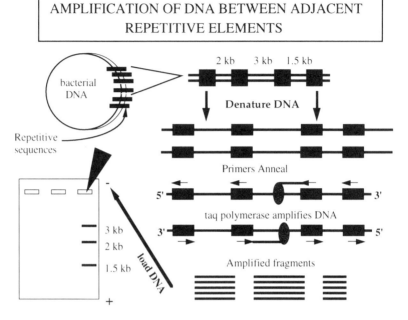

Figure 2 The rep-PCR protocol. An expanded view of the chromosomal DNA (top right) highlights four repetitive sequences with an intervening, 3-, 2-, and 1.5-kb expanse of anonymous DNA. Employing a thermocycler, the DNA is denatured (94°C) and primers allowed to anneal at a reduced temperature. The annealing temperature is determined based on the primer sequence(s) used (see Table 1). The thermostable enzyme Taq polymerase recognizes the portion of double-stranded DNA and from the 3′ end of opposing primers, proceeds to synthesize complementary DNA between two repetitive sequences. The denaturation, annealing, and extension steps are repeated thirty to thirty-five times. The result of this amplification is the generation of a collection of DNA fragments of characteristic lengths, reflecting the chromosomal organization of a given bacterial strain, which can be separated using standard electrophoresis methods. In this simple example the profile comprises three PCR products of sizes 3.0, 2, and 1.5 kb, corresponding to the distance between adjacent primer annealing sites.

PCR pattern of a particular collection of strain(s) can be obtained in fourteen to twenty-seven hours.

TEMPLATE DNA FOR rep-PCR GENOMIC FINGERPRINTING

Strains for rep-PCR analysis can originate from culture collections or natural isolates, such as those obtained from soil, water, or plant materials [40a]. Pure cultures are generally essential for proper interpretation of rep-PCR patterns. Identical fingerprint patterns can be obtained from DNA isolated from microbes, whole cells collected from solid or liquid

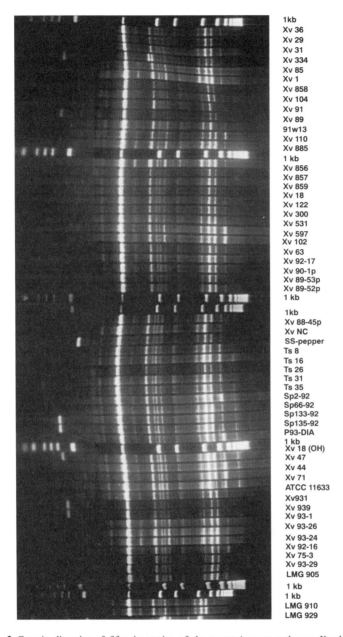

1kb
Xv 36
Xv 29
Xv 31
Xv 334
Xv 85
Xv 1
Xv 858
Xv 104
Xv 91
Xv 89
91w13
Xv 110
Xv 885
1 kb
Xv 856
Xv 857
Xv 859
Xv 18
Xv 122
Xv 300
Xv 531
Xv 597
Xv 102
Xv 63
Xv 92-17
Xv 90-1p
Xv 89-53p
Xv 89-52p
1 kb

1kb
Xv 88-45p
Xv NC
SS-pepper
Ts 8
Ts 16
Ts 26
Ts 31
Ts 35
Sp2-92
Sp66-92
Sp133-92
Sp135-92
P93-DIA
1 kb
Xv 18 (OH)
Xv 47
Xv 44
Xv 71
ATCC 11633
Xv931
Xv 939
Xv 93-1
Xv 93-26
Xv 93-24
Xv 92-16
Xv 75-3
Xv 93-29
LMG 905
1 kb
1 kb
LMG 910
LMG 929

Figure 3 Genetic diversity of fifty-six strains of the tomato/pepper pathogen *Xanthomonas campestris* pv. *vesicatoria* Group A collected from various regions of the world (see Table 1 in Reference [10]). The DNA extracted using the CTAB protocol was subject to rep-PCR using REP primers, and fingerprints were resolved on 1.5% agarose gels in 0.5× TAE. Fingerprint patterns are analogous to bar codes, such as UPC symbols in grocery stores, and can function as strain-specific signatures for diagnosis.

media, and from whole cells extracted directly from niches, such as plant lesions or root nodules [9,40a,41] (see also Figure 6). Either a large-scale or a small-scale protocol is used in order to obtain genomic DNA for rep-PCR fingerprinting. The large-scale preparation is based on the cetyltrimethylammonium-bromide (CTAB) procedure for Gram-negative bacteria [42], with a slight modification for Gram-positive bacteria.

LARGE SCALE GENOMIC DNA ISOLATION

Materials and supplies are as follows:
- Luria broth (LB) [43]
- NBY (nutrient broth 8 g/l; yeast extract 2 g/l; K_2HPO_4 2 g/l; KH_2PO_4 0.5 g/l)
- TY ($CaCl_2$ 0.37 g/l; tryptone 5 g/l; yeast extract 3 g/l; pH 6.8)
- 50 ml Oakridge tubes
- clinical centrifuge
- 1 M NaCl
- 0.1 M phosphate buffer, pH 6.8
- TEE (10 mM Tris pH 8.0; 25 mM EDTA)
- mutanolysin (Sigma)
- lysozyme
- achromopeptidase
- 10% Sarkosyl in TEE
- 5 mg/ml pronase
- 10% CTAB solution (100 mg CTAB in 1 ml 1 M NaCl per sample)
- 5 M NaCl
- chloroform
- phenol
- 10-ml blunt-end pipette (made by filing the tips off of disposable glass pipette)
- isopropanol
- 70% ethanol
- TE (10 mM Tris pH 8.0, 1 mM EDTA)
- Pasteur pipette (flamed to form a shepherd's crook)

Initiate a 40-ml liquid culture from a single bacterial colony and allow growth for one to two days at optimum temperature in a shaker (~ 200 rpm). The liquid media used should be low in carbohydrates in order to minimize polysaccharide production. Luria broth (LB) or NBY is used for *Xanthomonas* and *Clavibacter* or TY for *Rhizobium* species.

Streak a drop of the bacterial culture on general laboratory media (LB or NBY) with an inoculating loop to obtain single colonies and evaluate resultant growth after several days to ensure that no contamination occurred.

Pour the 40-ml culture into 50-ml Oakridge tubes and spin in a clinical centrifuge (maximum speed) for ten minutes. Pour off the supernatant or aspirate it if the pellet is loose.

For Gram-negative bacteria, resuspend the cells in 5 ml 1 M NaCl to wash the cells. Spin and discard the supernatant. Repeat the NaCl wash. For Gram-positive bacteria, wash the cells 2 times in 1 M NaCl, as described above. Wash a third time using 5 ml of 0.1 M phosphate buffer, pH 6.8. Freeze the cells at $-20°C$ until after the purity check is complete or follow the directions in the next paragraph. For Gram-negative bacteria, resuspend the cells in 5 ml TEE. In the case of Gram-positive bacteria, resuspend the cells in 5 ml 0.1 M phosphate buffer pH 6.8 and add mutanolysin (Sigma M9901) to a final concentration of 10 units/ml; incubate the mixture for a half hour at 37°C. The use of 10 mg/ml each of lysozyme and achromopeptidase has been found to be useful instead of mutanolysin.

To both Gram-negative and Gram-positive cell preparations, add 1 ml 10% Sarkosyl in TEE and 1 ml 5 mg/ml pronase that has been predigested in 0.1 M phosphate buffer (pH 7.0) for one hour at 37°C [43]. Mix and incubate two hours at 37°C. During this incubation period, prepare a 10% CTAB solution, and allow it to dissolve at 65°C.

Add 1.7 ml 5 M NaCl and 1 ml 10% CTAB. Mix well and incubate at 65°C for ten minutes. An optional additional step involves the addition of RNAse to a final concentration of 20 μg/ml and subsequent incubation of the mixture at 55°C for ten minutes. Removal of RNA facilitates the quantitation of DNA and may limit competition for primers, but this step is not essential for successful rep-PCR fingerprinting.

Add 5 ml chloroform and mix well. Spin the solution ten minutes in a clinical centrifuge. If the phases do not partition well (i.e., if there is excess white precipitate), spin in a Sorval centrifuge at 10,000 rpm for fifteen minutes. Alternatively, add more phosphate buffer so that more DNA will enter in the aqueous phase. Remove the supernatant with a 10-ml, blunt-end pipette and transfer it to a second Oakridge tube.

Add 2.5 ml each of chloroform and phenol (saturated in TE) and extract as described above. Repeat this procedure one to three more times, until there is no more white precipitate at the interface of the organic solvent and aqueous phases.

Transfer the aqueous phase to a fresh tube and add 0.6 × volumes of cold isopropanol; incubate on ice for ten minutes.

Spool out the DNA with a Pasteur pipette and wash the DNA with 70% ethanol. Resuspend the DNA in 1 ml TE or sterile, double-distilled water and determine the DNA concentration [43].

SMALL-SCALE GENOMIC DNA ISOLATION

Materials and supplies are as follows:

- microcentrifuge
- 1.5-ml microcentrifuge tubes
- TE, pH 8.0 (10 mM Tris pH 8.0, 1 mM EDTA)
- lysozyme
- achromopeptidase
- 5% SDS
- 5 mg/ml pronase
- phenol
- phenol:chloroform (1:1)
- absolute ethanol or isopropanol
- 70% ethanol

Collect 3 ml of a well-grown liquid culture by centrifugation in a 1.5-ml microcentrifuge tube ($\times$ 2 centrifugations), or collect cells from one or two Petri dishes bearing confluent growth. Resuspend the cells in 1.8 ml TE, pH 8.0, and pellet the cells again via centrifugation.

Resuspend the pellet in 0.3 ml TE, pH 8.0. At this step, for Gram-positive strains, 10 mg/ml each of lysozyme and achromopeptidase have been successfully included, followed by incubation for a minimum of a half hour at 37°C.

Add 100 μl 5% SDS to the cell mixtures and 100 μl predigested pronase (5 mg/ml; see above). Incubate the mixture at 37°C for one hour.

Shear the cells by pipetting the mixture up and down using a 1000-μl tip, until a foamy liquid is generated. Extract the mixture once with an equal volume of saturated phenol and twice with saturated phenol:chloroform (1:1).

Precipitate the DNA with 2 volumes of absolute ethanol or 1-volume isopropanol. Mix and centrifuge at room temperature for ten minutes.

Wash the pellet with approximately 0.5 ml 70% ethanol, and resuspend the nucleic acids in 200 μl TE, pH 8 (or water) and measure the DNA concentration.

RAPID TEMPLATE ISOLATION FROM ROOT NODULES OR FRUIT LESIONS

Materials and supplies are as follows:

- legume root nodules
- polypropylene grinder
- 1.5-ml microcentrifuge tubes
- microcentrifuge
- extraction buffer (200 mM Tris HCL, pH 7.5, 250 mM NaCl; 25 mM EDTA and 0.5% SDS [44]
- isopropanol

Collect one nodule or large lesion (e.g., from tomato fruit), sterilize the surface, and grind by hand with a polypropylene grinder in a dry, 1.5-ml microcentrifuge tube at room temperature.

Add 400 μl extraction buffer and vortex for fifteen seconds. Allow the mixture to stand five minutes or until all samples are prepared and centrifuge them at 13,000 rpm for three minutes.

Transfer 300 μl of the supernatant to a clean microcentrifuge tube and add 300 μl isopropanol. Let the mixture stand at room temperature for ten minutes. Centrifuge at 13,000 rpm for five minutes, pour off the supernatant, air-dry the pellet, and dissolve it in 100 μl H_2O or TE.

Determine the concentration of DNA (as indicated above), adjust to 150 ng/μl and use 1 μl for rep-PCR fingerprinting. Plant tissue known to be bacteria-free should be prepared in a similar manner as a control.

STORAGE OF ISOLATED DNA

Concentrated original solutions are stored at $-20\,^{\circ}C$ and working solutions of 50 ng/μl are stored at $4\,^{\circ}C$. Working solutions have been successfully stored over two years and used for rep-PCR fingerprinting.

WHOLE CELL rep-PCR FINGERPRINTING AND MODIFICATIONS

Whole cells can be taken from liquid medium, solid medium, or directly from plant lesions.

Materials and supplies are as follows:

- plates with isolated colonies, liquid cultures, legume root nodules or plant lesions
- 1.5-ml microcentrifuge tubes
- microcentrifuge
- 1 M NaCl
- 1-μl inoculating loops (Lab Product Sales, Rochester, NY)
- 70% ethanol
- 10% bleach (0.525% active hypochlorite)
- sterile distilled water
- 0.1 M phosphate buffer (pH 7.0) or 0.01 M $MgSO_4$ buffer (pH 7.0)

(1) Liquid cultures: take 3 ml of liquid culture (OD600 0.65–0.95) and spin down the cells for approximately five minutes into a microcentrifuge tube (two centrifugations). Wash the pellet with 1 M NaCl and spin the cells again for about four minutes. Strains that produce excess polysaccharides benefit from one to two additional washes with 1 M NaCl. Resuspend the pellet in 100 μl H_2O and store aliquots at $-20\,^{\circ}C$. Use 1–2 μl per 25 μl rep-PCR mix. The use of liquid cul-

tures directly without a washing step has been successful, but extracellular components often interfere with DNA amplification.

(2) In cells harvested directly from solid media, success has been achieved employing whole-cell rep-PCR to generate total genomic fingerprints from colonies of *Rhizobium* sp., *Clavibacter michiganensis* subsp., *E. coli,* various pseudomonads and xanthomonads as well as from a large collection of unidentified subsurface microbes. Success with direct collection of whole-cells from solid medium depends on the bacterial species, the medium supporting bacterial growth, and the number of cells collected. Successful results have been obtained with plate cultures of selected strains up to four weeks old, but in general more reliable fingerprint patterns are generated from fresh colonies [40a].

In research using serial dilutions of bacteria from soil or plant samples a minimum of four to six days of growth (25–28°C) was allowed before single colonies were selected for whole-cell PCR. Although colonies may be visible sooner and subject to amplification for a more rapid diagnosis, too many nontarget colonies are inadvertently selected. Therefore, it is better to wait until the colony morphology can be used as a criterion to counter select "suspect colony forming units" for rep-PCR. Optimum waiting periods should be established and are dependent on the growth rate of the strain, the semi-selective media used, and the urgency of the diagnosis.

Care should be taken to avoid collecting too many cells from colonies on plates. Excess cell numbers result in poor amplification of the DNA or in poor visualization of the amplification products (smears on agarose gels). The best results are obtained by using disposable 1-μl inoculating loops. An entire small single colony ($<$ 1 mm diameter) can be collected or the loop is passed along the edge of larger colonies. Excess cells can be removed by drawing the loop on the surface of the medium where there is no growth. The loop is then placed into microfuge tubes containing 24 μl of the PCR mixture and cells are dislodged by rotating the loop or by flicking the loop gently. Samples can also be successfully collected using plastic pipetter tips, but the use of toothpicks is discouraged since soluble ingredients in the wood may interfere with the Taq polymerase enzyme activity [45].

Other rapid protocols have been used to prepare crude extracts of DNA useful for rep-PCR fingerprinting. Versalovic et al. [11] described a preheating step for samples that include *Bacillus subtilis* and *Pseudomonas aeruginosa.* Samples containing cells, buffer, and H_2O are heated to 80°C in the thermal cycler. After fifteen minutes, tubes are kept at 80°C, and dNTPs, primers, and Taq enzyme are added followed by the cycling steps outlined below. In addition to preheating, sonication has also proven useful [6].

Plant nodules and lesions as sources of bacterial template DNA: nodule extracts from legume plants have yielded rep-PCR fingerprints identical to purified genomic DNA [40a]. Likewise, plant lesions, if collected at the right stage of infection can contain a near-pure culture of the pathogen from which diagnostic rep-PCR fingerprints can be generated. Success has been achieved generating rep-PCR fingerprints from cells collected from *Pseudomonas* and *Xanthomonas* infected tomato foliar lesions; *Xanthomonas* lesions on geranium leaves [9] and three-day-old, as well as dried one-year-old *Xanthomonas* lesions on rice leaves [41].

For plant lesions, the lesion and surrounding tissue are excised and surface sterilized in 70% ethanol for one minute and 10–20% bleach for one minute.

The tissue is rinsed 3 times in sterile distilled water and then sliced into small pieces in the presence of 200 μl 0.1 M phosphate buffer (pH 7.0), 0.01 M $MgSO_4$ buffer (pH 7.0; preferred for xanthomonads) or H_2O. Allow the preparation to stand for fifteen to thirty minutes to enable the bacteria to dislodge from the dissected tissue.

Collect 1 μl of the solution, taking care to include as little plant material as possible, and dispense the cells in 24 μl of rep-PCR reaction mixture. Contaminating plant material and excess phenolics or other contaminating materials may interfere with successful PCR amplification. Therefore, do not macerate the tissue more than necessary.

DETAILED rep-PCR PROTOCOL

STOCK SOLUTIONS AND WORKING SOLUTIONS

The preparation of stock solutions and PCR reaction mixtures is outlined below. Special care must be used to avoid contamination of the PCR ingredients. Stock solutions are prepared from molecular biology-grade ingredients and/or are autoclaved.

5× GITSCHIER BUFFER [46]

Prepare and autoclave stock solutions of each ingredient, and subsequently combine them to prepare a 5× buffer. For 200 ml of 5× buffer, proportion the stock solutions to achieve a final concentration of each of the following ingredients using sterile, double-distilled water:

- 83 mM NH_4SO_4 (16.6 ml of a 1 M stock)
- 335 mM Tris HCl pH 8.8 (67 ml of a 1 M stock)
- 33.5 mM $MgCl_2$ (6.7 ml of a 1 M stock)
- 33.5 μM EDTA (1.3 ml of 1:100 dilution of a 0.5 M stock)

- 150 mM β-mercaptoethanol (2.08 ml of a 14.4 M commercial stock)

Dispense buffer into 1.5-ml microcentrifuge tubes and store at $-20\,^{\circ}$C. This buffer can be stored for several months.

BSA

Bovine serum albumin (BSA) stabilizes enzymes, such as Taq polymerase and should be nuclease-free (e.g., use Boehringer #711 454 molecular biology–grade BSA, 20 mg/ml). Aliquot the BSA in 20-μl amounts and freeze at $-20\,^{\circ}$C.

DMSO

Dimethyl sulfoxide allows for reduced annealing temperatures [47] and is added to the PCR mixture at a final concentration of 10% v/v. FLUKA (#41640) can be used and 1-ml aliquots in 1.5-ml microcentrifuge tubes are stored at $-20\,^{\circ}$C.

dNTPs

2$'$-deoxynucleoside-5$'$-triphosphates are the building blocks for the newly synthesized DNA. Pharmacia #27-2035-01 ultrapure set of 100 mM in H_2O, pH 7.5 is used. Mix each dNTP 1:1:1:1 (working solution of 25 mM of each dNTP) and store aliquots at $-20\,^{\circ}$C.

Taq POLYMERASE

Primarily, AmpliTaq DNA polymerase at 5 units/μl from Perkin-Elmer #N808-0070 is used. The enzyme is aliquoted in 10-μl amounts and stored (up to several months) at $-20\,^{\circ}$C.

TWEEN 20

Tween 20 in some cases enhances whole-cell lysis and the quality of the rep-PCR fingerprints generated. Approximately 0.5 ml is dispensed into plastic, screw-cap vials and autoclaved. Tween 20 is added to a final concentration of 1% v/v in the 25-μl reaction mixture.

MINERAL OIL

Sigma M3516 light, white mineral oil, with preparation similar to Tween 20 above, is used to overlay the PCR reaction mixture and to prevent evaporation.

WATER

The HPLC purified water (e.g., Fisher, Pittsburgh, PA) can be used to ensure the absence of contaminating DNA. However, double-distilled sterile water is generally used and dispensed as 2.5-ml aliquots in 5-ml, screw-cap vials and autoclaved.

PRIMERS

The DNA sequences of the primers used are shown in Table 1. Standard PCR primers can be synthesized in-house or obtained commercially. The desiccated preparation is resuspended in 200 μl H_2O and a 5-μl primer solution is mixed with 395-μl H_2O to determine its concentration. Measure the absorbance at 260 nm using a spectrophotometer and calculate the concentration of the original 200-μl sample as outlined [43].

Rep-PCR AMPLIFICATION OF GENOMIC DNA

Materials and supplies are as follows:

- rep-PCR mastermix ingredients
- template DNA
- microcentrifuge
- thermocycler

Prepare a master mix of 24 μl for each sample that contains 5 μl 5× Gitschier buffer, 0.2 μl BSA, 2.5 μl DMSO, an appropriate amount of double-distilled sterile water, 1.25 μl of 1:1:1:1 dNTP solution, 1 μl of each primer solution (e.g., 1 μl each of REP1RI & REP2-I or ERIC1R & ERIC2 or 1 μl BOXA1R), 2 units of Taq polymerase, and an optional 0.25 μl of Tween 20. Mix by inverting several times or if Tween 20 is included, briefly vortex the master mix to ensure that the ingredients are well mixed. The master mix can be prepared on ice in a sterile 1.5-ml microcentrifuge tube or a 5-ml glass vial and then dispensed.

To each sample add 1 μl of genomic DNA or whole cell preparation and overlay the reaction mixtures with an equal volume of mineral oil. Spin the samples briefly in a microcentrifuge to ensure that the reaction mixture is submerged under the oil and that air bubbles are removed.

Load the samples in the thermocycler and use the following cycling regimes:

- one initial cycle at 95°C for seven minutes
- 30–35 cycles at 94°C for one minute; annealing as outlined in Table 1 for one minute and extension at 65°C for eight minutes

- a single final extension cycle at 65°C for fifteen minutes
- final soak at 4°C

For purified DNA of good quality, thirty cycles are sufficient to generate quality rep-PCR fingerprint patterns. For crude DNA or whole-cell rep-PCR use thirty-five cycles. Completed PCR reactions from purified DNA have been successfully stored at 4°C for over two years with no apparent DNA degradation. Whole-cell PCR reactions have also been stored successfully at 4°C, but with certain preparations degradation has been observed and therefore samples are best stored at −20°C.

SEPARATION OF rep-PCR PRODUCTS

The rep-PCR fragments can be successfully resolved using agarose-based matrices, although acrylamide-based systems can also be employed (not described here).

AGAROSE GEL ELECTROPHORESIS

Rapid systems, such as a 1% minigel [11], for resolving DNA fingerprints are useful if a simple diagnosis is desired. However, for detailed comparisons and analyses, considerably better resolution of the amplified DNA bands is achieved using 20 cm × 24 cm gels (Figures 3–7), for example, by using a BRL H4 horizontal gel apparatus (#1025RD). Thirty- or twenty-tooth combs can be used and the wells should be thin (1 mm) to ensure that the resolved rep-PCR bands are sharp.

Solutions and Materials

50× TAE buffer stock solution includes 242 g/l Tris Base, 57.1 ml/l glacial acetic acid, and 18.61 g/l EDTA disodium salt. Autoclave and store at room temperature. Use 10 ml/l to prepare a 0.5× TAE running buffer.

5× loading dye: Bromophenol blue 0.25% w/v in dH_2O; xylenecyanol 0.25% w/v and Ficoll 15% w/v Type 400 [43].

A 1-kb ladder (Gibco BRL) can be used successfully as a DNA marker. Alternatively, the combined markers Lambda Hind III/pUc 19/Taq I-pUC 19/Sau3AI (Stratagene CAT#201113) provide a series of well-spaced markers within the size range of the amplified rep-PCR fragments (see Figure 7).

A 1.5% agarose gel is optimal for resolution of DNA fragments in the 0.2–3 kb range [43].

Gel Loading, Electrophoresis, and Data Recording

Place the gel in its gel chamber and cover it with running buffer. Mix 6 μl PCR mixture with 1.2 μl of loading dye on a parafilm strip and load the samples in the well. To prepare the size markers, combine 2 μl of a 1 kb ladder solution and mix with 4 μl 0.5 × TAE and 1.2 μl loading dye. At a minimum, load the DNA marker in the two outside and one middle well of each gel [Figures 3–4(a)], especially if any type of automated post-electrophoresis analysis is to be performed. Gels can be run for five to six hours at a constant voltage of 105 V (approximately 55 mA) at room temperature or for sixteen to eighteen hours at a constant voltage of 70 V (approximately 23 mA) and 4 °C for optimal results (Figures 3–4).

After electrophoresis, transfer the gels to an ethidium bromide bath (60 μl of 10 mg/ml stock solution per one L of TAE buffer) for a half-hour and destain in 0.5 × TAE buffer for a half-hour. Photograph the gel on an ultraviolet transilluminator, using an orange UV filter and positive film (e.g., Polaroid type 57) or film that provides a negative and positive (e.g., Polaroid type 55). Alternatively, gel images can be captured electronically with a video recorder and saved as a TIFF file by using the appropriate software.

ANALYSIS OF rep-PCR FINGERPRINTS

Data analysis of the genomic fingerprints, especially in studies that include a large number of diverse strains, can be a challenge without some sort of image analysis software tool and supporting hardware.

SCORING BY EYE

If fingerprint patterns yield limited polymorphisms among strains (e.g., Figure 3) or if the number of strains that are being compared is limited, the total number of scorable bands for each strain can be determined, and the rep-PCR bands can be scored as present (1) or absent (0) for each unique position across all strains (see References [4,9] for examples). A binary-code rectangular matrix is created on a spreadsheet and can be imported into software, such as NTSYS-PC [48] (Applied Biostatistics, Inc., Setauket, NY). A similarity matrix is created using the subroutine SIMQUAL (similarity for qualitative data), employing one of various algorithms, and a dendrogram is generated based on the similarity matrix using the subroutine SAHN (sequential, agglomerative, hierarchical, and nested or nonoverlapping clustering methods). For example, the UPGMA

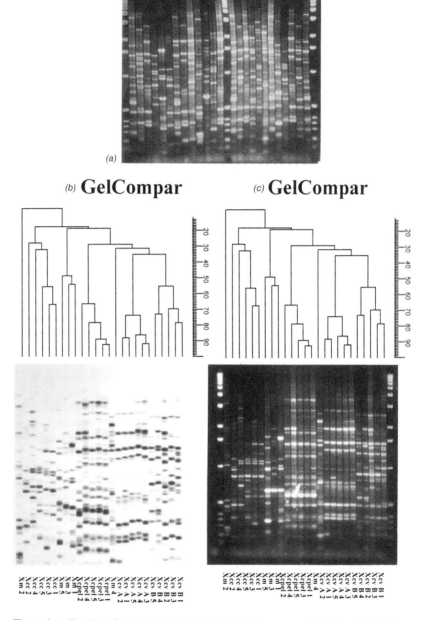

(b) GelCompar **(c) GelCompar**

Figure 4 (a) The DNA from twenty-five xanthomonad strains was amplified using ERIC primers and samples loaded in an unorganized manner. The gel image was scanned into GelCompar and normalized based on the three reference standards (1 kb ladder). Thresholds were set and strain-specific patterns were matched using Pearson's correlation coefficient and the dendogram constructed using UPGMA. (b) GelCompar constructed the resultant dendogram and displayed rear-

(unweighted, pair-group method with arithmetic averages) can be used for cluster analysis and a measure of fit of the dendrogram(s) generated can be determined using the Cophenetic values subroutine.

COMPUTER-BASED ANALYSIS OF rep-PCR FINGERPRINTS

Often, large numbers of bacteria need to be compared and in many cases the strains to be compared comprise considerable diversity in rep-PCR fingerprint patterns [e.g., Figure 4(a)]. For these studies, computer-assisted pattern analysis is essential. Several hardware/software combinations are available, and those that have been most useful will be highlighted.

Computer-Assisted Analysis of Standard Agarose-Resolved Patterns

Agarose gels are loaded, as a minimum [Figure 4(a)], with DNA markers in the two outermost lanes and one in the center. Additional marker lanes can be used to further minimize intragel variations of DNA bands. The DNA molecular weight markers are essential to normalize gels to account for variable rates of migration and minimize inter- and intragel variations. Each gel also should include an internal standard (positive control; see details below) used to set the threshold for determining the scorability of bands and for determining the appropriate level of smoothing required for resolving the rep-PCR bands/peaks.

One available commercial software package that has been used for analysis of rep-PCR fingerprint patterns is the IBM-based AMBIS system. The use of AMBIS and examples has been described in detail [11,40a]. Original images are recorded with a video camera and captured for electronic storage, using the AMBIS MicroPM software. The region of interest is defined: parameters are set including background "noise" to be subtracted, band-recognition thresholds based on the intensity required for a band to be detected, and a filter parameter that alters a smoothing function for resolving peaks proximal to one another. The sample lanes are delineated and named, and each gel is normalized to an absolute standard [11].

Samples from several gels, or individual sample lanes from different gels, can be grouped and stored in a band base. The band base is then used to generate a similarity matrix using various algorithms, such as Simple,

ranged images of the lane profiles for rapid evaluation of clustering results. (c) A second gel using the original PCR mixtures was loaded according to the order specified by GelCompar. The Xcv B, Xcv A, Xcpel and Xcc are *Xanthomonas campestris* pv. *vesicatoria* Group B, Group A (tomato/pepper pathogens), pathovar *pelergonii* (geranium pathogen) and pathovar *campestris* (crucifer pathogen) respectively. The Xm is *X. maltophilia* an ecologically diverse species.

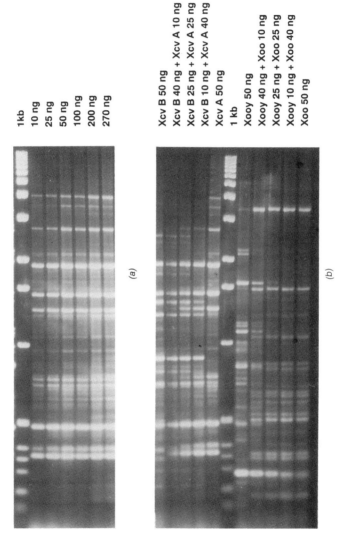

1kb
10 ng
25 ng
50 ng
100 ng
200 ng
270 ng

(a)

Xcv B 50 ng
Xcv B 40 ng + Xcv A 10 ng
Xcv B 25 ng + Xcv A 25 ng
Xcv B 10 ng + Xcv A 40 ng
Xcv A 50 ng
1 kb
Xooy 50 ng
Xooy 40 ng + Xoo 10 ng
Xooy 25 ng + Xoo 25 ng
Xooy 10 ng + Xoo 40 ng
Xoo 50 ng

(b)

Figure 5 (a) Effect of increasing concentration of template DNA using ERIC primers. The DNA was extracted using the CTAB protocol and quantified spectrophotometrically. Dilutions were prepared and 1 μl of each dilution was added to 24 μl of PCR reaction mixture. Samples were amplified and resolved on a 1.5% agarose gel. (b) Effect of mixing different proportions of DNA from two strains that may be able to occupy the same niche (plant lesion or tissue). The DNA was prepared as in 1 above, and samples were mixed in various proportions. Subsequently, 1 μl of each mixture was added to PCR reaction mixtures, amplified, and resolved.

Pattern matching, or Jaccard coefficients. The matrix is used to generate a dendrogram using UPGMA to highlight the scored similarity among strains. A second set of gels can be loaded in the order specified by AMBIS to facilitate visual comparison of fingerprint patterns [11,49].

A second commercial software package is GelCompar (Applied Maths, Belgium). Gels should be loaded with DNA molecular markers and internal standards, as described above for analysis with the AMBIS system. The gel image from a scanned photo or TIFF file is imported [Figure 4(a)]. GelCompar features automated or manual lane-finding capabilities and detailed associated sample information. Nodes can be added to account for "snaking" lanes, in order to highlight all the data associated with the samples. Gels are normalized according to standard reference patterns and/or by aligning internal reference peaks. Background subtraction and thresholds are defined and up to 1000 patterns may be clustered using one of several coefficients and clustering algorithms. Furthermore, lane tracks (images) can be electronically rearranged and displayed for easy viewing and visual comparison of rep-PCR fingerprint profiles [e.g., Figure 4(c)], thus omitting the reloading step required when using the AMBIS system [40a]. GelCompar also features a library manager and identification module to compare unknown patterns to those in a database.

CONSIDERATIONS FOR OPTIMIZING rep-PCR AND COMPARISON TO OTHER PROTOCOLS FOR ASSESSING MICROBIAL DIVERSITY

rep-PCR AND RAPD ANALYSIS

The rep-PCR primers are designed to anneal with medium to high stringency to specific sequences distributed throughout the chromosome of diverse bacteria. While the possibility that primers fortuitously anneal to nonrepetitive sequences cannot be ruled out, experience has shown that rep-PCR, especially with ERIC and BOX primers, is quite robust and not subject to the extent of variability reported for other PCR-based fingerprinting protocols, such as the RAPD technique [50]. However, in the case of the RAPD method, reproducible fingerprint patterns and reliable information about diversity within microbial communities can also be obtained employing optimized and uniform amplification conditions [21,23,24]. Similar to RAPDs, knowledge of specific DNA sequences is not required to generate rep-PCR fingerprint patterns. However, a uniform set of primer(s) and PCR conditions is used to generate rep-PCR fingerprint patterns from a wide range of bacterial species, whereas optimum primers and PCR conditions must be empirically determined with the RAPD method. The RAPD technique generally includes annealing temperatures from 35°C to 38°C to encourage annealing of the nine- to eleven-base arbitrary primers and the subsequent amplification of random DNA fragments. Temperatures above 40°C can limit amplification [21]. The rep-PCR employs more stringent annealing temperatures and longer primers (Table 1). In the case of REP primers, a lower annealing temperature of 40°C to 44°C is used to compensate for the presence of inosine as a substitute base. However, even under such conditions rep-PCR fingerprint patterns are reproducible (Figure 3).

An arbitrary primer in the RAPD protocol may yield four to twenty informative DNA fragments per genome. In contrast, each rep-PCR primer set can yield up to fifteen or more DNA fragments providing a number of scorable characters to discern strain diversity (Figures 3–4) [2,4,9,10].

SOURCE OF DNA

The rep-PCR is not highly sensitive to DNA concentration [Figure 5(a)] and perhaps this facilitates whole-cell PCR and reproducibility of fingerprint patterns. This observation has also provided considerable flexibility in the need to carefully quantify DNA preparations.

The recommendation to isolate DNA or use whole-cell rep-PCR depends on the population under study and the objective of the study. Bacteria that produce excess polysaccharides or that lyse with difficulty may not yield dependable results using whole-cell PCR. In such cases, a DNA extraction protocol, as described above and elsewhere in this volume, may prove beneficial. Also, if the objective is to perform a detailed genetic analysis of a large population employing several primer sets, DNA isolation will facilitate reproducibility and limit the number of individual samples that yield poor patterns. For example, to analyze a collection of over 150 xanthomonad strains, we used isolated DNA [9,10]. Xanthomonads produce polysaccharides that can limit efficient amplification of rep-PCR profiles. In contrast, whole-cell PCR worked efficiently for the analysis of a collection of over 300 *Clavibacter* strains and a second collection of over seventy *E. coli* strains (Louws, Hausbeck, Fulbright and de Bruijn, unpublished data). To determine the most efficient approach, it is recommended that a subsample of strains be included in a large study to compare the efficiency of rep-PCR using whole cells or isolated DNA. Success with whole-cell rep-PCR may also depend on the media used for bacterial growth (Figure 6).

Pure cultures are essential to determine the true identity of a strain since mixed cultures will yield mixed non-stoichiometric patterns [Figure 5(b)] [40a]. However, the fact that rep-PCR generates bands from each strain present in a mix may be used to an advantage. For example, two distinct strains (Xcv A and Xcv B) cause a similar disease on tomato foliage and

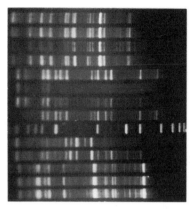

CMM 936 DIRECT KB
CMM 936 DIRECT NYDA
CMM 936 DIRECT SCM2-t
CMM 936 DNA
XCV 982 DIRECT KB
XCV 982 DIRECT NYDA
XCV 982 TWEEN B
XCV 736 DNA
1 kb LADDER
UNKNOWN DIRECT FROM LESION
UNKNOWN DIRECT FROM LESION
PST 915 DIRECT KB
PST 915 DIRECT NYDA
PST 915 DNA

Figure 6 Effect of whole-cell PCR for different phytopathogenic bacteria and effect of medium from which whole cells were collected. *Clavibacter michiganensis* subsp. *michiganensis (CMM)*, *Xanthomonas campestris* pv. *vesicatoria* (Xcv), and *Pseudomonas syringae* pv. tomato (PST) were initiated on King's B (KB), nutrient yeast dextrose agar (NYDA), SCM2-t (semiselective for CMM), or Tween B (semiselective for XCV). After seven days of growth, a small number of cells were collected and placed directly into microfuge tubes containing reagents for PCR and primers corresponding to boxA. Samples were subject to conditions outlined in this chapter. Samples of tomato from a commercial field with a presumptive bacterial epidemic were brought to the lab. Lesions were cut from fruit and a DNA miniprep was performed. One 1 μl of extract was subjected to PCR as described. The DNA extracted from isolates proven to be true pathogens by Koch's postulates were used as positive controls (DNA). Many thanks to Nancy Fichter for preparing this and similar gels.

fruit [10], and two distinct strains (Xoo and Xooy) may be found on a single leaf of rice [51]. Figure 5(b) demonstrates that a simple community analysis is possible when two strains occupy the same plant lesion or were extracted from the same plant tissue.

PREPARATION OF PCR REACTION MIXTURES

The protocols outlined in this chapter have been used successfully by numerous laboratories for fingerprinting diverse strains of bacteria, but the protocols should be carefully followed in order to be successful, especially in case of initial studies. In many cases when rep-PCR has not worked in other laboratories, the problem could be traced to the 5× Gitschier buffer. Proprietary and other buffers sold with Taq polymerase enzymes have not always proven to be the most efficient. Moreover, the best results are obtained by autoclaving individual stock solutions and subsequently preparing the 5× buffer as outlined. The BSA should be added just prior to preparing sample master mixes or precipitation may result. Prolonged storage of the 5× buffer may result in the formation of a white precipitate, but upon heating, the suspension is redissolved with no apparent effect on rep-PCR patterns.

Stock solutions of primers, enzyme, and dNTPs are aliquoted using filtered tips and/or a biological UV hood to prevent contamination. However, for preparation of sample master mixes and addition of template DNA, standard tips and dispensors on an open bench are used. Surface sterilization of the working area and use of a Bunsen burner with an open flame may help to limit contamination of sample tubes during preparation. Alternatively, use of positive displacement dispensors, separate areas for handling bacterial cultures, and preparation of PCR reaction mixtures, as well as the use of gloves, further aid in the limitation of sample cross contamination (see Chapter 2).

Each experiment should include a control, lacking template DNA, to ensure that no bands are amplified due to contamination of stock solutions and other materials used, and a positive control, usually purified DNA from a well-characterized strain, to ensure that the PCR reaction worked successfully. The positive control should preferably generate a pattern of PCR-amplified bands that range in size from less than 500 bp to over 3500 bp. If the larger bands are not amplified or if a normal intensity of the bands is not apparent, then the PCR run was suboptimal and may need to be repeated. Also, if the positive control is included on each agarose gel, profiles can be normalized for computer assisted analysis.

CHOICE OF PRIMER(S)

For detailed analyses of a microbial community, combined data from REP-, ERIC-, and BOX-PCR provide a more detailed assessment of chromosomal organization and diversity among strains than single fingerprints. Each primer set is equally effective in differentiating strains and each primer set may offer unique information among closely related strains. The most informative primer for a given set of strains depends on the microbial populations to be examined and the objective (i.e., to find similarities or differences among strains). For example, the Xcc strains #1–5 in Figure 4(c) share few comigrating bands using ERIC-PCR whereas BOX-PCR and REP-PCR generate more bands that comigrate [9].

If the goal is to differentiate strains or rapidly identify an unknown strain compared to a known control, BOX-PCR has proven the most useful. BOX-PCR appears to be less sensitive to contaminants in DNA or whole cell preparations and therefore generates interpretable patterns even under suboptimal conditions.

It is possible to prepare the master mix with or without template DNA and store the PCR reaction mixtures in the freezer until the time is convenient to put the tubes in the thermocycler (Figure 7).

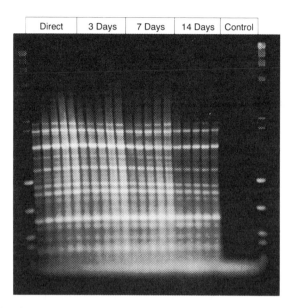

| Direct | 3 Days | 7 Days | 14 Days | Control |

Figure 7 Effect of storing prepared master mixes including template DNA. Stored for zero, three, seven and fourteen days at −20°C prior to amplification.

CALIBRATION OF rep-PCR

The rep-PCR results have been compared to results obtained using other protocols, and such studies have provided considerable confidence in the utility of rep-PCR for assessing diversity among closely related strains and for determining the population structure of microbial communities. Rep-PCR differentiates serotypes of *Bradyrhizobium* and provides more strain-specific profiles [4]. Data generated using rep-PCR is consistent with results obtained by RFLP analysis [4,41]. In the latter case, DNA from thirty-one *Xanthomonas oryzae* pv. oryzae (Xoo) strains, a serious pathogen of rice, was sent to the laboratory for analysis as part of a "blind" study. The REP-, ERIC-, and BOX-PCR were performed and strains were compared and grouped according to their rep-PCR patterns using various clustering coefficients and algorithms. The combined data set of rep-PCR patterns detected polymorphisms among Xoo strains and arranged genetic groups in a manner consistent with those determined by RFLP analysis [41].

Also, rep-PCR was found able to group strains in clusters in a manner consistent with the relationships determined by multilocus enzyme electrophoresis (MLEE) [2, Louws and de Bruijn, unpublished]. Efforts to calibrate rep-PCR suggest rep-PCR should prove to be a useful tool, complementary to other techniques, for the rapid determination of genetic diversity among closely related strains within microbial communities.

INTERPRETING POLYMORPHISMS

Polymorphisms among closely related strains can arise by several mechanisms, including changes in the DNA sequence of annealing sites, deletions, or inversions. Therefore, polymorphisms that arise from the loss of bands are much more likely between two strains as compared to two strains acquiring homologous rep-PCR bands. Therefore, rep-PCR is useful for comparison of closely related strains when the diversity in banding patterns is low. In cases where few bands comigrate, conclusions concerning phylogenic relationships are more complex.

APPLICATIONS OF rep-PCR

Rep-PCR is proving useful for many applications in the analysis of microbial communities. A few specific cases are highlighted.

CLASSIFICATION

Rep-PCR has proven useful for classifying bacteria of medical importance [3,7,13] and for ecological/agricultural experimental purposes [10,52,53]. Environmental studies of important microbes often require an efficient means to cluster strains within groups. Once grouped, representative strains can be selected for further studies. For example, Ka et al. [54] used rep-PCR to classify strains able to degrade 2,4-D isolated from agricultural plots treated with 2,4-D. The rep-PCR analysis complemented other genotypic, chemical, and physiological methods to rapidly classify isolated strains into four groups and provided useful data concerning the genetic diversity within each group.

Rep-PCR has also proven useful as a classification tool for taxonomic purposes, to group closely related strains, as determined by other polyphasic taxonomical criteria, or to differentiate strains that cannot easily be differentiated using other techniques [2,4,9,10]. For example, *Bradyrhizobium japonicum* is an indigenous symbiont in the Midwest, U.S.A., which nodulates soybeans. A collection of strains has been classified through serology as serocluster 123, but the serological divisions do not adequately reflect the genetic and phenotypic diversity of the member isolates [4]. However, rep-PCR effectively differentiated, among these closely related strains, in contrast to other classification protocols, and therefore proved to be a useful tool to examine intraserogroup competition among serocluster 123 strains and their ability to nodulate soybeans [4].

Moreover, rep-PCR effectively differentiates pathogenic variants (pathovars) of xanthomonads and pseudomonads that cannot be reliably differentiated through traditional phenotypic and biochemical characteristics [9]. For example, ERIC-PCR generates pathovar or strain-specific profiles among various xanthomonads (Figure 4).

The rep-PCR can differentiate genetic groups and provide information about diversity within each group (of closely related strains), but generally is not effective for discerning relatedness among strains from genetically different groups. Other genetic tools, such as rDNA-based techniques, are required for such phylogenetic information.

DIAGNOSTICS

The rep-PCR technique has proven useful for diagnosing bacteria of medical importance [8,14] and for ecological/agricultural experimental purposes [2,9,10]. The value of rep-PCR for diagnostics depends on the population structure of the microbial community and the number of known fingerprint profiles determined for that community. For example, Xcv A is comprised of a homogeneous population distributed world-wide (Figure 3) [10]. The observed fingerprint is highly diagnostic for this tomato/pepper pathogen. Also, Xcpel appears to be comprised of a homogeneous population [Figure 4(c)]. Xcv B is comprised of distinct lineages based on several ERIC-PCR polymorphisms but strains share numerous diagnostic features [Figure 4(c)] [10]. In contrast, Xcc appears to be comprised of a more heterogeneous collection of strains [Figure 4(c)] [9] and fingerprint patterns are less diagnostic.

In extreme cases, strains classified within the same taxon may be genetically diverse, based on rep-PCR analysis. For example, Xm includes strains that are ecologically diverse, and in our example three Xm strains clustered together and the other two strains were intermingled in this limited database of twenty-five xanthomonads [Figure 4(c)]. No obvious ERIC-PCR diagnostic feature is observed for these five strains. Such observed diversity may be an issue of unresolved classification considerations or may reflect the rapid evolution of chromosomal organization among strains within a species. In any case, the population structure of a microbial community must be determined in order to know the potential value of rep-PCR as a diagnostic tool.

EPIDEMIOLOGICAL ANALYSES

Strain-specific profiles enable one to monitor or select specific strains in epidemiological studies. The rep-PCR has proven useful for clinical applications in order to determine the origin of pathogenic strains and the vertical transmissibility of strains [5,7,14]. The presence of polymorphisms among closely related strains in order to design ecological and epidemiological experiments of symbiotic and plant pathogenic bacteria have been exploited.

SUMMARY

The rep-PCR fingerprinting approach is a versatile tool that has proven useful for classification and diagnostic and epidemiological studies in medical, industrial, ecological, and agricultural microbiology. The rep-PCR fingerprinting complements other approaches to classify and assess diversity among microbial populations and provides information about the genetic diversity at the species, subspecies or strain levels. With rep-PCR, a unified set of primers, corresponding to repetitive DNA sequences and standardized PCR conditions, can be used to generate complex fingerprint patterns from a wide range of bacterial species. The protocol is relatively rapid and requires a modest investment for equipment and supplies. Fingerprinting bacteria using repetitive-based PCR will likely continue to develop as an important tool in molecular characterization of microbes. Future advances will be driven by the need for a more detailed understanding of microbial communities and population dynamics and by advances in PCR-based technologies, such as automated systems for generating and analyzing rep-PCR fingerprint patterns.

REFERENCES

1 Versalovic, J., T. Koeuth, and J. R. Lupski. 1991. Distribution of repetitive DNA sequences in eubacteria and application to fingerprinting of bacterial genomes. *Nucleic Acids Res.* 19:6823–6831.

2 de Bruijn, F. J. 1992. Use of repetitive (repetitive extragenic palindromic and enterobacterial repetitive intergeneric consensus) sequences and the polymerase chain reaction to fingerprint the genomes of *Rhizobium meliloti* isolates and other soil bacteria. *Appl. Environ. Microbiol.* 58:2180–2187.

3 Woods, C. R., J. Versalovic, T. Koueth, and J. R. Lupski. 1992. Analysis of relationships among isolates of *Citrobacter diversus* by using DNA fingerprints generated by repetitive sequence-based primers in the polymerase chain reaction. *J. Clin. Microbiol.* 30:2921–2929.

4 Judd, A. K., M. Schneider, M. J. Sadowsky, and F. J. de Bruijn. 1993. Use of repetitive sequences and the polymerase chain reaction technique to classify genetically related *Bradyrhizobium japonicum* serocluster 123 strains. *Appl. Environ. Microbiol.* 59:1702–1708.

5 Lupski, J. R. 1993. Molecular epidemiology and its clinical application. *JAMA.* 270:1363–1364.

6 Woods, C. R., J. Versalovic, T. Koueth, and J. R. Lupski. 1993. Whole-cell repet-

itive element sequence-based polymerase chain reaction allows rapid assessment of clonal relationships of bacterial isolates. *J. Clin. Microbiol.* 31:1927–1931.

7 Versalovic, J., C. R. Woods, Jr., P. R. Georghiou, R. J. Hamill, and J. R. Lupski. 1993. DNA-based identification and epidemiologic typing of bacterial pathogens. *Arch. Pathol. Lab. Med.* 117:1088–1098.

8 Versalovic, J., V. Kapur, E. O. Mason, Jr., U. Shah, T. Koeuth, J. R. Lupski, and J. M. Musser. 1993. Penicillin-resistant *Streptococcus pneumoniae* strains recovered in Houston: Identification and molecular characterization of multiple clones. *J. Infect. Dis.* 167:850–856.

9 Louws, F. J., D. W. Fulbright, C. T. Stephens, and F. J. de Bruijn. 1994. Specific genomic fingerprints of phytopathogenic *Xanthomonas* and *Pseudomonas* pathovars and strains generated with repetitive sequences and PCR. *Appl. Environ. Microbiol.* 60:2286–2295.

10 Louws, F. J., D. W. Fulbright, C. T. Stephens, and F. J. de Bruijn. 1995. Differentiation of genomic structure by rep-PCR fingerprinting to rapidly classify *Xanthomonas campestris* pv. *vesicatoria. Phytopathology.* 85:528–536.

11 Versalovic, J., M. Schneider, F. J. de Bruijn, and J. R. Lupski. 1994. Genomic fingerprinting of bacteria using repetitive sequence-based polymerase chain reaction. *Meth. Mol. Cell. Biol.* 5:25–40.

12 Lupski, J. R. and G. M. Weinstock. 1992. Short, interspersed repetitive DNA sequences in prokaryotic genomes. *J. Bacteriol.* 174:4525–4529.

13 Georghiou, P. R., A. M. Doggett, M. A. Kielhofner, J. E. Stout, D. A. Watson, J. R. Lupski, and R. J. Jamill. 1994. Molecular fingerprinting of *Legionella* species by repetitive element PCR. *J. Clin. Microbiol.* 32:2989–2994.

14 Georghiou, P. R., R. J. Hamil, C. E. Wright, J. Versalivic, T. Koeuth, D. A. Watson, and J. R. Lupski. 1995. Molecular epidemiology of infections due to *Enterobacter aerogenes:* Identification of hospital outbreak-associated strains by molecular techniques. *Clin. Inf. Diseases.* 20:84–94.

15 Koeuth, T., J. Versalovic, and J. R. Lupski. 1995. Interspersed repetitive *S. pneumoniae* BOX elements contain subsequences which are conserved in diverse bacteria. Submitted for publication.

16 Wayne, L. G., et al. 1987. Report of the *ad hoc* committee on reconciliation of approaches to bacterial systematics. *Int. J. Sys. Bacteriol.* 37:463–464.

17 Schmidt, T. M. 1994. Fingerprinting bacterial genomes using ribosomal RNA genes and operons. *Meth. Mol. Cell. Biol.* 5:3–12.

18 Massol-Deya, A. A., D. A. Odelson, R. F. Hickey, and J. M. Tiedje. 1995. Bacterial community fingerprinting of amplified 16S and 16-23S ribosomal DNA gene sequences and restriction endonuclease analysis (ARDRA). In: A. D. L. Akkermans, J. D. van Elsas and F. J. de Bruijn, (eds.), *Molecular Microbial Ecology Manual.* Kluwer Academic Publishers, Dordrecht. p. 3.3.2:1–8.

19 Jensen, M. A., J. A. Webster, and N. Straus. 1993. Rapid identification of bacteria on the basis of polymerase chain reaction-amplified ribosomal DNA spacer polymorphisms. *Appl. Environ. Microbiol.* 59:945–952.

20 Welsh, J. and M. McClelland. 1991. Genomic fingerprints produced by PCR with consensus tRNA gene primers. *Nucleic Acids Res.* 19:861–866.

21 Williams, J. G. K., A. R. Kubelik, K. J. Livak, J. A. Rafalski, and S. V. Tingey. 1990. DNA polymorphisms amplified by arbitrary primers are useful as genetic markers. *Nucleic Acids Res.* 18:6531–6535.

22 Welsh, J. and M. McClelland. 1990. Fingerprinting genomes using PCR with arbitrary primers. *Nucleic Acids Res.* 18:7213–7218.

23 Bowditch, B. M., D. G. Albright, J. K. Williams, and M. J. Braun. 1993. Use of randomly amplified polymorphic DNA markers in comparative genome studies. In: *Molecular evolution: Producing the biochemical data. Meth. Enzym. 224.* Zimmer, E. A., T. J. White, R. L. Cann, and A. C. Wilson, (eds.) Academic Press, Inc. NY.

24 Berg, D. E., N. S. Akopyants, and D. Kersulyte. 1994. Fingerprinting microbial genomes using the RAPD or AP-PCR method. *Meth. Mol. Cell. Biol.* 5:13–24.

25 Weising, K., H. Nybom, K. Wolff, and W. Meyer. 1995. *DNA fingerprinting in plants and fungi.* CRC Press, Inc. Ann Arbor, MI.

26 Parish, J. E. and D. L. Nelson. 1994. Practical aspects of fingerprinting human DNA using *Alu* polymerase chain reaction. *Meth. Mol. Cell. Biol.* 5:71–77.

27 Krawiec, S. 1985. Minireview: Concept of a bacterial species. *Int. J. Syst. Bacteriol.* 35:217–220.

28 Krawiec, S. and M. Riley. 1990. Organization of the bacterial chromosome. *Microbiol. Rev.* 54:502–539.

29 Gilson, E., D. Perrin, J. M. Clément, S. Szmelcman, E. Dassa, and M. Hofnung. 1986. Palindromic units from *E. coli* as binding sites for a chromoid-associated protein. *FEBS.* 206:323–328.

30 Newbury, S. F., N. H. Smith, E. C. Robinson, I. D. Hiles, and C. F. Higgins. 1987. Stabilization of translationally active mRNA by prokaryotic REP sequences. *Cell.* 48:297–310.

31 Higgins, C. F., R. S. McLaren, and S. F. Newbury. 1988. Repetitive extragenic palindromic sequences, mRNA stability and gene expression: Evolution by gene conversion: A review. *Gene.* 72:3–14.

32 Yang, Y. and G. F-L. Ames. 1988. DNA gyrase binds to the family of prokaryotic repetitive extragenic palindromic sequences. *Proc. Natl. Acad. Sci. USA.* 85:8850–8854.

33 Gilson, E., D. Perrin, and M. Hofnung. 1990. DNA polymerase I and a protein-complex bind specifically to *E. coli* palindromic unit highly repetitive DNA: Implications for bacterial chromosome organization. *Nucleic Acids Res.* 18:3941–3952.

34 Dimri, G. P., K. E. Rudd, M. K. Morgan, H. Bayatt, and G. F-L. Ames. 1992. Physical mapping of repetitive extragenic palindromic sequences in *Escherichia coli* and phylogenetic distribution among *Escherichia coli* strains and other enteric bacteria. *J. Bacteriol.* 174:4583–4593.

35 Higgins, C. F., G. F. Ames, W. M. Barnes, J. M. Clement, and M. Hofnung. 1982. A novel intercistronic regulatory element of prokaryotic operons. *Nature.* 298:760–762.

36 Stern, M. J., G. F. L. Ames, N. H. Smith, E. C. Robinson, and C. F. Higgins. 1984. Repetitive extragenic palindromic sequences: a major component of the bacterial genome. *Cell.* 37:1015–1026.

37 Hulton, C. S. J., C. F. Higgins, and P. M. Sharp. 1991. ERIC sequences: a novel family of repetitive elements in the genomes of *Escherichia coli, Salmonella typhimurium* and other enterobacteria. *Mol. Microbiol.* 5:825–834.

38 Sharples, G. J. and R. G. Lloyd. 1990. A novel repeated DNA sequence located in the intergenic regions of bacterial chromosomes. *Nucleic Acids Res.* 18:6503–6508.

39 Martin, B., O. Humbert, M. Camara, E. Guenzi, J. Walker, T. Mitchell, P. Andrew, M. Pruhdomme, G. Alloing, R. Hakenbeck, D. A. Morrison, G. J. Boulnois, and J.-P. Claverys. 1992. A highly conserved repeated DNA element located in the chromosome of *Streptococcus pneumoniae*. *Nucleic Acids Res.* 20:3479-3483.

40 Garnber, H. R. Automating the PCR process. In: K. B. Mullis, F. Ferré, R. A. Gibbs, (eds.) *The polymerase chain reaction*. Birkhäuser, Boston. pp. 182-198.

40a Schneider, M. and F. J. de Bruijn. 1996. Rep-PCR mediated genomic fingerprinting of rhizobia and computer assisted phylogenetic pattern analysis. *World J. Microbiol. Biotechnol.* In Press.

41 Vera Cruz, N., L. Halda, F. J. Louws, D. Z. Skinner, M. L. George, R. J. Nelson, F. J. de Bruijn, C. Rice, and J. E. Leach. 1995. Repetitive sequence-based PCR of *Xanthomonas oryzae* pv. *oryzae* and *Pseudomonas* species. *Int. Rice Res. Notes.* 20:23.

42 Ausubel, F. M., R. Brent, R. E. Kingston, D. D. Moore, J. G. Seidman, J. A. Smith, and K. Struhl. 1992. *Current protocols in molecular biology. Vol. I.* Greene Publishing Assoc. and Wiley-Interscience, NY.

43 Sambrook, J., E. F. Fritsch, and T. Maniatis. 1989. *Molecular cloning: A laboratory manual,* 2nd ed. Cold Spring Harbor Laboratory Press, NY.

44 Edwards, K., C. Johnstone, and C. Thompson. 1991. A simple and rapid method for the preparation of plant genomic DNA for PCR analyses. *Nucleic Acids Res.* 19:1349.

45 Lee, A. B. and T. A. Cooper. 1995. Improved direct PCR screen for bacterial colonies: Wooden toothpicks inhibit PCR amplification. *BioTechniques.* 18:225-226.

46 Kogan, S., M. Doherty, and J. Gitschier. 1987. An improved method for prenatal diagnosis of genetic diseases by analysis of amplified DNA sequences. *N. Engl. J. Med.* 317:985-990.

47 Pomp, D. and J. F. Medrano. 1991. Organic solvents as facilitators of polymerase chain reaction. *BioTechniques.* 10:58-59.

48 Rohlf, F. J. 1992. NTSYS-PC: Numerical taxonomy and multivariate analysis system. Version 1.70. Exeter Software, Setauket, NY.

49 Rossbach, S., D. A. Kulpa, U. Rossbah, and F. J. de Bruijn. 1994. Molecular and genetic characterization of the rhizopine catabolism (*moc*ABRC) genes of *Rhizobium meliloti* L5-30. *Mol. Gen. Genet.* 245:11-24.

50 Ellsworth, D. L., K. D. Rittenhause, and R. L. Honeycut. 1993. Artificial variation in randomly amplified polymorphic DNA banding pattern. *BioTechniques.* 14:214-217.

51 Mew, T. W., A. M. Alvarez, J. E. Leach, and J. Swings. 1993. Focus on bacterial blight. *Plant Dis.* 77:5-12.

52 Leung, K., S. R. Strain, F. J. de Bruijn, and P. J. Bottomley. 1994. Genotypic and phenotypic comparisons of chromosomal types within an indigenous soil population of *Rhizobium leguminosarum* bv. *trifolii. Appl. Environ. Microbiol.* 60:416-426.

53 Strain, S. R., K. Leung, T. S. Whittam, F. J. de Bruijn, and P. J. Bottomley. 1994. Genetic structure of *Rhizobium leguminosarum* biovar trifolii and viciae populations found in two Oregon soils under different plant communities. *Appl. Environ. Microbiol.* 60:2772-2778.

54 Ka, J. O., W. E. Holben, and J. M. Tiedje. 1994. Genetic and phenotypic diversity of 2,4-dichlorophenoxyacetic acid (2,4-D)-degrading bacteria isolated from 2,4-D-treated field soils. *App. Environ. Microbiol.* 60:1106–1115.

55 Selander, R. K. and J. M. Musser. 1990. Population genetics of bacterial pathogenesis, pp. 11–36. In: B. H. Iglewski and V. L. Clark (eds.), *Molecular basis of bacterial pathogenesis.* Academic Press, Inc., San Diego, CA.

CHAPTER 4

Detection of Microorganisms in Soils and Sludges

IAN L. PEPPER[1]
TIMOTHY M. STRAUB[1]
CHARLES P. GERBA[1]

BACTERIA

APPROACHES FOR PROCESSING SOIL AND SLUDGES FOR PCR AMPLIFICATIONS OF BACTERIAL NUCLEIC ACIDS

THE PCR has been successfully used to detect bacterial DNA and RNA in soil and sludges, but most work has been conducted on DNA. In soil the target organisms have usually been pathogens, indicators of pathogens, or organisms capable of degradation of xenobiotics including genetically engineered bacteria. Thus, many bacterial primers exist for the detection of specific organisms. Pathogenic bacteria are often introduced into soil, either deliberately when sewage, sludge, or animal wastes are disposed of in soil, or through opportunistic mechanisms, such as bird droppings. It is, of course, well documented that sewage and sludges can contain specific bacterial pathogens, such as *Salmonella* spp. or viral pathogens including enteroviruses [1]. Such organisms can survive days, weeks, or even months depending on soil moisture and temperature [1,2]. Thus, there is the potential for disease incidence, which makes detection of pathogens in soil and sludges critical.

Steffan and Atlas [3] conducted one of the earliest PCR studies on soil using target DNA from *Pseudomonas* spp. Fecal coliforms in soil have also been detected via PCR [4]. Pillai et al. [5] amplified Tn5 specific sequences within *Rhizobia* spp. extracted from soil. Other examples include multiplex amplification of *Salmonella* spp. [6] and plasmic pJP4 contained

---

[1]Department of Soil, Water and Environmental Science, 429 Shantz Building #38, College of Agriculture, The University of Arizona, Tucson, AZ 85721, U.S.A.

within *Alcaligenes eutrophus* [7], which codes for the degradation of 2,4-dichlorophenoxyacetic acid. In sludges, most PCR studies have centered on the detection of indicators or pathogens. Tsai and Olson [8] developed a method for direct detection of *E. coli* in sludge, whereas Way et al. [6] detected *Salmonella* in a municipal waste pond.

Thus, there is ample evidence for the successful amplification of DNA sequences in soil and sludges, although the process is not without its difficulties. The PCR amplification requires prior extraction and purification of bacterial DNA from the environmental sample. There have been two approaches to obtaining the DNA. In one approach, whole bacterial cells are separated from colloidal material, purified, and subsequently lysed prior to analysis [3,4,5]. This method suffers from selective extraction of different genera of bacteria and has generally been superseded by the *in situ* direct-lysis method. Here, cells are lysed directly in the soil, followed by extraction and purification of the DNA [8,9].

Although mRNA is difficult to extract from environmental samples, there have been a few successful studies. Tsai et al. [19] utilized guanidine thiocyanate to successfully extract mRNA from soil. More recently Moran et al. [10] successfully extracted rRNA from soil, which, after purification, was sufficiently pure to allow subsequent molecular analyses.

PROTOCOLS FOR PCR AMPLIFICATION IN SOIL OR SLUDGE

Three protocols are outlined:

(1) Extraction of cells using $CaCl_2$ and sucrose density centrifugation followed by cell lysis [5]
(2) In situ cell lysis followed by DNA extraction [8]
(3) Purification of community DNA

Extraction of Cells Using $CaCl_2$ and Sucrose Density Centrifugation Procedure [5]

Special Apparatus

(1) Sorvall SS-34 rotor
(2) Beckman J-6B centrifuge
(3) A DNA temperature cycler

Reagents

(1) $CaCl_2$ (1%) sterile solution prepared using anhydrous $CaCl_2$ pellets
(2) Sucrose solution (1.33 g/ml); solution prepared by dissolving 855 g of

sucrose in 450 ml or water; solution sterilized by autoclaving at 121 °C for five minutes

(3) A PCR amplification kit

Procedures—Extraction of Cells:

(1) Nine milliliters of sterile 1% $CaCl_2$ is added to 1 g of soil or sludge subsamples containing variable number of cells. The soil-$CaCl_2$ slurry is vigorously vortexed for one minute and then allowed to settle for one hour.

(2) The upper fraction is concentrated to 5 ml by centrifugation (7650 g, five minutes) and pipetted into a 50 ml polypropylene centrifuge tube (Becton Dickinson, NJ) containing 10 ml of (1.33 g/ml) sucrose.

(3) The soil-cell sucrose slurry is vigorously vortexed for one minute. Fifteen milliliters of (1.33 g/ml) sucrose is carefully layered underneath the slurry.

(4) This biphasic gradient is centrifuged at 750 g for ten minutes.

(5) The clear dilute sucrose upper fraction (~ 12 ml) that contains the bacterial cells is pipetted into a 40-ml-polyallomer centrifuge tube, diluted with water, and centrifuged at 12,000 g for twenty minutes at 10°C in a SS-34 rotor.

(6) The bacterial cells, along with environmental colloids that had a similar density to the cells, appear as a pellet. This pellet is directly used as the template for the PCR amplifications.

Double PCR Amplification

(1) The pellet obtained after the extraction process contains bacterial cells, soil colloids, and perhaps free, naked DNA adsorbed to colloids.

(2) The pellet is resuspended in 30 μl of H_2O and added to the PCR reaction mix containing the reaction buffer, nucleotides (Perkin-Elmer Cetus, Norwalk, CT), and primers.

(3) The reaction mix is heated to 98°C in the DNA thermal cycler (Perkin-Elmer Cetus, Norwalk, CT) for ten minutes to lyse the cells, and subsequently cooled briefly to room temperature.

(4) The *AmpliTaq* DNA polymerase (Perkin-Elmer Cetus) is added and twenty-five cycles of PCR performed using the DNA thermal cycler.

(5) At the end of twenty-five cycles, 10 μl of the amplified product is added to a fresh reaction mix and further amplified for twenty-five cycles, thus performing a "double" PCR protocol.

(6) The amplified products are detected using agarose gel electrophoresis or specific gene probes (see Chapters 2 and 3).

In Situ Cell Lysis Followed by DNA Extraction [8]

Special Apparatus

(1) Centrifuge
(2) DNA temperature cycler

Reagents

(1) Sodium phosphate buffer: 120 mM at pH 8.0 Lysis Solution: 0.15 m NaCl, 0.1 M Na_2 EDTA, pH 8.0 containing 15 mg of lysozyme/ml.
(2) NaCl: 0.1 M
(3) Tris-HCl: 0.5 M, pH 8.0
(4) Sodium dodecyl sulfate: 10%
(5) Tris HCl: 0.1 M, pH 8.0, saturated with phenol
(6) Chloroform mixture: chloroform/isoamyl alcohol, 24:1
(7) Isopropanol
(8) TE buffer: 20 mM Tris HCl, 1 mM EDTA, pH 8.0
(9) Pancreatic RNAse A: final concentration 0.2 $\mu g/\mu l$

Procedures

(1) Soil or sludge (1 g sample) is mixed with 2 ml of sodium phosphate buffer and shaken at 150 rpm for fifteen minutes. The slurry is pelleted by centrifugation at 6000 × g for ten minutes.
(2) The pellet is washed with phosphate buffer and then resuspended in 2 ml of lysis solution prior to incubation in a 37°C water bath for two hours with agitation every thirty minutes. Two ml of the NaCl/Tris HCl/SDS mixture are then added.
(3) Three cycles of freezing in a −70°C dry ice/ethanol bath and thawing in a 65°C water bath are conducted to release DNA from bacterial cells.
(4) Two ml of Tris HCl saturated phenol is added and the sample emulsified by briefly vortexing.
(5) The mixture is centrifuged at 6000 × g for ten minutes. A ml of the top aqueous layer is collected and mixed with 1.5 ml of phenol and 1.5 ml of chloroform. A 2.5-ml portion of the resulting extract is further extracted with an equal volume of the chloroform mixture.

(6) Nucleic acids in the extracted aqueous phase (2 ml) are precipitated with 2 ml of cold isopropanol at $-20°C$ for one hour or overnight.

(7) The crude nucleic acid pellet is harvested by centrifugation at 10000 × g for ten minutes, then vacuum dried at 23°C.

(8) The pellet is resuspended in 100 μl of TE buffer, and RNA is removed by incubation with heat-treated pancreatic RNase A.

(9) The RNA-free DNA is then purified by passage through an Elutip-d column (Schleicher and Schuell, Keene, NH) attached to a Schleicher and Schuell NA010/27 (0.45 μm cellulose acetate) prefilter. The DNA is recovered from the column, as suggested by the manufacturer.

(10) The DNA can now be PCR amplified and detected as described previously.

Purification of Community DNA

Community DNA extracted from soil or sludge is often subject to PCR inhibition via humic contaminants. Such contaminated DNA can be purified by a variety of filtration techniques including spun columns of DNA purification kits. Spun columns of Sephadex G-200 (Pharmacia, 2KB Biotechnology Inc., Piscataway, NJ) or Elutip-d columns (Schleicher and Schuell, Keene, NH) have been popular choices for the purification of DNA.

SPECIFIC PRIMERS FOR ORGANISMS IN SLUDGE

(1) Fecal Coliforms   Gene = *lamB*

5′	TCA	CTg	CTg	AAC	ATA	CTC	AgA	gT	3′
5′	gTC	gAg	gAT	ACg	CAg	CAT	gTg	AC	3′

(2) *Salmonella*   Multiplex PCR

phoP	5′	A T g C A A A g C C C g A C C A T g A C g	3′
	5′	g T A T C g A C C A C C A C g A T g g T T	3′
Hin	5′	C T A g T g C A A A T T g T g A C C g C A	3′
	5′	C C C C A T C g C g C T A C T g g T A T C	3′
H-li	5′	A g C C T C g g C T A C T g g T C T T g	3′
	5′	C C g C A g C A A g A g T C A C C T C A	3′

PITFALLS AND PROBLEMS

Sorption of DNA by Colloidal Material

It is known that colloidal material has the potential to irreversibly sorb

DNA of bacterial origin [11]. This is particularly prevalent in fine-textured soils or soils high in organic material. Sludges also contain high organic contents capable of sorption. If DNA harvests are particularly low, sorption of DNA may be the problem. At this point, there is no effective strategy to overcome this problem.

Lack of Amplification

A negative PCR result may not necessarily mean that the target organism or gene sequence is not present. To check for the presence of a false negative, environmental samples should be seeded with the target gene sequence, the sample extracted, and PCR amplified to ensure that PCR is not being inhibited. The level of seeding should not be excessive. For whole cells, perhaps 10^3 CFU/g soil or sludge should be added. If inhibition is occurring, samples should be treated, for example, by passage through Elutip columns of Sephadex G-200. For positive and negative controls see Chapter 2.

Sensitivity of Detection

This is important when a given DNA sequence is only present in an environmental sample at low concentration, and lack of sensitivity may result in a false negative. Sensitivity can be evaluated in terms of whole-cell lysates or pure genomic DNA either of which can be seeded into soil or sludge. Sensitivity in terms of whole cells can be misleading since the total number of copies of a target sequence is always greater than the number of CFUs. Broth cultures inevitably contain dead or lysed cells with target sequences, and each viable cell is likely to contain multiple copies of each genome [12]. Therefore, equating sensitivity with CFUs tends to overestimate the actual sensitivity of the method. Sensitivity also depends on the method used to detect the amplified product. Using ethidium-bromide staining of DNA, PCR can often detect from 10^3 to 10^4 CFUs, whereas the use of ^{32}P-labelled gene probes usually increases the sensitivity by two orders of magnitude. Using PCR in soil, a sensitivity of one cell per gram of soil has been reported [3,5].

In terms of pure DNA, 100 ag of a 179-bp fragment has been amplified [4]. Assuming a total genome of 4×10^6 bp, which is equivalent to approximately 9 fg of DNA, the target amplification product of 179 bp is equivalent to $(179/4 \times 10^6) \times 9 \times 10^{-15}$ g or 0.4 ag. Therefore, one copy of the target DNA represents 0.4 ag and the sensitivity of detection in this study was approximately 250 copies (100 ag).

Overall, the issue of sensitivity is complex and must be evaluated for each set of primers. Such an evaluation is critical if PCR amplifications are to be used for diagnostic purposes.

Specificity of Amplification

The degree of specificity of primers used in PCR can be intentionally varied. If highly specific primers are required to amplify specific DNA, then unique DNA sequences must be chosen as the target for amplification, for example, a primer pair that allows only amplification of *Salmonella* spp.-specific DNA, but not *E. coli*-specific DNA. Since these two groups of organisms are closely related, the design of these primers is critical to distinguish between the species of the two genera. However, if the detection of all or most bacterial DNA in an environmental sample is required (e.g., in a bioassay), then primers termed "universal primers" are designed that allow amplification of a conserved DNA sequence present in all bacteria (Chapter 2).

The degree of specificity can be varied by primer design and also by changing the annealing temperature. In general, as the annealing temperature is raised, the number of base-pair mismatches allowable for hybridization decreases, increasing the specificity of amplification. Increasing the annealing temperature from 50°C to 55°C will often decrease nonspecific amplification. However, along with increased specificity, there is an associated decrease in sensitivity. The maximum allowable annealing temperature is generally 10°C less than the melting temperature that depends on GC%.

Primers can sometimes result in several different amplification products as well as an amplification product specific to the target DNA. The amplification product from the correct target sequence can be identified by appropriate size markers of standard DNA. The correct amplification product can also be identified by the use of end-labelled gene probes specific to an internal region of the amplified product. Nonspecific amplification products can arise in many ways. These include products that are due to the primers annealing to incorrect regions of the template, incomplete amplification, or truncated product formation. Such nonspecific products generally arise at low-annealing temperature conditions or when the extension time is not long enough.

Detection of Nonviable Bacteria

A positive amplification event implies that the target DNA sequence was present in the environmental sample. It does not, however, imply that viable bacteria were necessarily in the sample. Josephson et al. [4] showed PCR detection of nonviable bacterial pathogens that had been UV killed. However, in most environmental samples, such dead organisms normally degrade quickly at ambient temperatures. Interpretation of positive results should, therefore, be done carefully, depending on the state of the environmental sample.

APPLICATIONS IN ENVIRONMENTAL MICROBIOLOGY

Detection of gene sequences in soil or sludge by PCR is important for several reasons. It allows detection of specific orgasnisms, which include pathogens and biodegradative organisms of interest including *Alcaligenes* spp. [7]. It can also be used to estimate the potential for enzymatic activity by utilizing conserved primer sequences that code for specific activity [7]. The PCR amplifications of rDNA can also be used in evolutionary and biodiversity studies. Finally, note that PCR amplifications of soil community DNA can be used to estimate horizontal gene transfer and genomic rearrangements by utilizing primers designed from repetitive gene sequences.

VIRUSES

APPROACHES FOR PROCESSING SOIL AND SEWAGE SLUDGES FOR PCR AMPLIFICATION FOR VIRAL NUCLEIC ACIDS

The nucleic acids of viruses, unlike bacteria, are composed entirely of RNA or DNA. Most viruses of interest in sewage sludge and soil are the human enteric viruses, of which a majority are RNA viruses. Hence, normal DNA PCR cannot be performed directly. Instead, the RNA must first be converted to cDNA through reverse transcription, after which PCR can be performed. Inhibition from heavy metals and organics present in sludge and soil can prevent reverse transcription, PCR, or both.

Data on successful amplification of human enteric virus RNA in sewage sludge and sludge-amended soil is limited and mostly describes methods employed to remove inhibition from these samples. Ansari et al. [13] employed phenol chloroform extraction to remove PCR inhibition and concentrate viral nucleic acids in sewage sludge but reported that the method was unsuccessful in terms of lack of amplification compared to existing cell-culture data. Kopecta et al. [14] employed solvent extraction methods as well for activated sludge samples. Here, six of ten samples resulted in positive amplification. However, all ten samples were cell-culture positive.

Straub et al. [15] found that spun-column chromatography employing size exclusion and ion exchange chromatography successfully removed PCR inhibiting compounds from a variety of soil types amended with sewage sludge. This method was also applied to several undigested and anaerobically digested liquid-sludge samples. Amplification of enterovirus RNA was reported in eight of eight samples, and amplification of hepatitis A virus was reported in seven of eight samples [1]. Graff et al.

[16] employed "antigen-capture" techniques to successfully amplify hepatitis A virus RNA in sewage-sludge samples. Here, PCR tubes were coated with HAV antibody and with viruses potentially present, were allowed to adsorb to the antibodies. Inhibition was removed by a series of washes leaving the "antigens" captured.

While chromatography and antigen-capture methods appear to be successful in removing PCR inhibition, issues still remain with PCR amplification of enteric viruses from sewage sludge and soil samples including: (1) optimization of viral recovery from these samples, (2) successful sample concentration procedures to assay larger equivalent sample volumes/weights, and (3) viability.

PROTOCOLS FOR PCR AMPLIFICATION OF RNA HUMAN ENTERIC VIRUSES IN SLUDGE AND SOIL

The following protocols are outlined.

Viral Recovery from Sewage Sludge

(1) sewage sludge [17]
(2) sewage sludge [16]
(3) soil [15]

Purification for Sewage Sludge and Sludge-Amended Soil

(1) column-chromatography methods [1,15]
(2) antigen-capture methods [16]

RT-PCR

(1) direct PCR [15]
(2) antigen capture (HAV; [16])

Viral Recovery from Sewage Sludge

Sewage Sludge [17]

SPECIAL APPARATUS

(1) Beckman JA-10 rotor capable of speeds up to 16,000 × g
(2) High-speed centrifuge
(3) pH meter

(4) Vacuum filter funnel
(5) Vacuum filter flask
(6) Vacuum source
(7) Sonicator capable of 100 W output

REAGENTS

Sterile 3% buffered-beef extract solution:
(1) 30 g beef extract V; BBL
(2) 73 g Na_2HPO_4
(3) 1.1 g citric acid
(4) Dissolve in 500 ml distilled water
(5) Adjust pH to 7.2 ± 0.2 with 1 N HCl or 1 N NaOH as needed
(6) Bring to 1 l volume with distilled water
(7) Autoclave

Sterile 0.15 M Na_3HPO_3 buffer, pH 9.0 ± 0.2
Freon (1,1,2-trichlorotrifluoroethane)

PROCEDURES

(1) 500 ml of liquid sewage sludge (ca. 1%–2% solids) is adjusted to pH 3.5 ± 0.1 using 1 N HCl and aluminum chloride (final concentration of 5.0 × 10^{-4} M in sludge).
(2) Stir up thirty minutes at the reduced pH.
(3) Centrifuge at 15,300 × g for ten minutes.
(4) Decant supernatant and resuspend pellet in 500 ml of sterile 3% beef-extract solution.
(5) Elute resuspended pellet for thirty minutes by constant stirring and adjust pH to 7.2 ± 0.2 if necessary.
(6) Centrifuge at 15,300 × g for thirty minutes.
(7) Retain supernatant and discard the pellet.
(8) Prefilter supernatant through a Whatman #1 filter with the aid of vacuum filtration.
(9) Adjust pH of filtrate to pH 3.5 causing flocculation of the beef-extract proteins.
(10) Centrifuge at 15,000 × g for ten minutes.
(11) Retain pellet and dissolve in 30 ml of sterile 0.15 M phosphate buffer.
(12) Adjust pH to 7.2 ± 0.2 and extract the concentrate with an equal volume of Freon.

(13) Centrifuge at 11,500 × g for ten minutes to separate phases and carefully pipette the aqueous (top) phase into a sterile 30-ml plastic scintillation vial.

Sewage Sludge [16]

REAGENTS

(1) 30 g beef extract V, (BBL) in 100-ml distilled water
(2) Freon (1,1,2-trichlorotrifluoroethane)
(3) 1 N HCl or 1 N NaOH
(4) Mixed serum pool of antibodies against enteroviruses, except HAV (only for HAV isolation by cell culture)

PROCEDURES

(1) Ten ml of 30% beef extract are added to 100 ml of sewage sludge.
(2) Adjust pH of the suspension to 9.0 with 1 N NaOH and sonicate (100 W × two minutes) on ice.
(3) Centrifuge for fifteen minutes at 4,000 × g.
(4) Adjust supernatant to pH 7.2 add freon to a final volume of 10%.
(5) Agitate suspension for fifteen minutes at 4°C. Separate phases collecting the upper, aqueous phase.
(6) If cell culture isolation of HAV is required, mix processed sample with a serum pool containing antibodies against all enteroviruses except HAV.

Soil [1]

REAGENTS

Sterile 3% buffered beef-extract solution:
(1) 30 g beef extract V (BBL)
(2) 7.3 g Na_2HPO_4
(3) 1.1 g citric acid
(4) Dissolve in 500 ml distilled water
(5) Adjust pH to 7.2 ± 0.2 with 1 N HCl or 1 N NaOH as needed
(6) Bring to 1 l volume with distilled water
(7) Autoclave

PROCEDURES

(1) Elute viruses from soil using 3% buffered-beef extract in a ratio of 10 ml of beef extract per gram of soil [17] by constant mixing on a stir plate for thirty minutes.

(2) Separate soil from the supernatant by centrifugation for thirty minutes at 15,000 × g.

(3) Retain supernatant and remove PCR inhibiting substances using one of the methods described below.

Sample Purification for Sewage Sludge and Soil

Column Chromotagraphy Methods for Sewage Sludge and Soil [1,15]

REAGENTS

(1) Sephadex G-50 equilibrated overnight in Tris (10 mM Tris-HCl)-EDTA (1 mM)-NaCl (100 mM) buffer, pH 8.3. After equilibration, autoclave.

(2) Chelex 100 (Bio Rad; Richmond, CA)

PROCEDURES

(1) One ml sterile disposable tuberculin syringes are plugged with either sterile tissue or glass wool. Needles, if present, are removed and discarded appropriately.

(2) Sephadex G-50 spun columns are prepared by filling the syringe with slurry. The syringe column is placed inside a sterile 15-ml conical centrifuge tube, and centrifuged (400 × g), and refilled with slurry until the partially dehydrated beads occupy a volume of 1 ml.

(3) Chelex 100 spun columns are prepared in the same manner with the exception that the resin requires no equilibration in buffer; for example, it is loaded using a sterile metal spatula and centrifuged (400 × g) until the 1 ml volume is attained.

(4) Remove the excess buffer from the 15 ml conical centrifuge tube by means of a sterile pipette. Then, 50 μl of concentrate is micropipetted onto the top of the Sephadex G-50 column and the partially dehydrated bands are allowed to expand around the sample.

(5) Wait ten minutes (room temperature) and centrifuge (400 × g) for ten minutes in a tabletop centrifuge. Collect the sample liquid from the bottom of the centrifuge tube using a sterile micropipette.

(6) Load the treated sample onto a fresh column containing 1 ml of Chelex 100. Carefully stir the sample into the top of the resin bed and allow to stand for thirty minutes at 4°C.

(7) Centrifuge (400 × g) for ten minutes. The liquid should be PCR compatible after this step. If not, repeat entire process with the treated sample.

Antigen Capture Methods, HAV [16]

REAGENTS

(1) Monoclonal antibodies to hepatitis A virus [16]

(2) 1% bovine serum albumin (BSA) solution

(3) Phosphate buffered saline solution containing 0.05% Tween 80

PROCEDURES

(1) Coat PCR tubes with monoclonal antibody using an optimized concentration determined for the stock antibody solution.

(2) Block antibodies with a 1% solution of BSA.

(3) Wash three times with PBS containing 0.05% Tween 80.

(4) Load 80 μl of sample to the coated PCR tubes and incubate at 4°C for a minimum of twelve hours.

(5) Wash samples five times with buffer containing 20 mM Tris-HCl (pH 8.4), 2.5 mM $MgCl_2$, 75 mM KCl. Washes eliminate/dilute inhibitory substances present in the sample.

(6) Perform RT-PCR

RT-PCR

Reagents

(1) GeneAmp RNA PCR Kit (Perkin-Elmer)

(2) RNase inhibitor (Promega)

(3) Random hexamers (Applied Biosystems)

(4) PCR Kit (Perkin-Elmer)

Direct PCR [15]

PROCEDURES

(1) Ten μl of the purified sample is incubated at 99°C for five minutes in

reaction buffer and nucleotides (Perkin-Elmer, Norwalk, CT) to release viral nucleic acids from the capsids.

(2) The reaction is cooled to 4°C and reverse transcriptase, as well as the RNAse inhibitor (Promega) and random primers are added. Samples are incubated at 25°C for ten minutes, 42°C for fifty minutes, 99°C for five minutes and then cooled to 4°C.

(3) Specific primers, reaction buffer, and *AmpliTaq* are added and 30 cycles of PCR are performed using a DNA thermal cycler.

(4) At the end of 30 cycles, 1 μl of the amplified product is added to a fresh reaction mix and further amplified for 30 cycles, thus performing "double" PCR, if the same primer set is used, or "seminested'" PCR, if at least one primer is internal to the amplified sequence.

Antigen Capture PCR [16]

REAGENTS

(1) Beef extract
(2) 10 mM Tris-HCl (pH 8.3)
(3) 50 mM KCl
(4) 1 mM MgCl$_2$
(5) dNTPs
(6) Avian myeloblastosis virus (AMV) reverse transcriptase (Promega)
(7) RNAsin (Promega)

PROCEDURES

(1) After the last wash to remove inhibitors, single-tube RT and PCR is performed. Denaturation is carried out by incubation of the sample at 95°C for five minutes in 10 mM Tris-HCl (pH 8.3), 50 mM KCl, 1.5 mM MgCl$_2$, 0.25 mM each dNTPs, 100 nM sense primer and 100 nM antisense primer.

(2) After denaturation, 5U AMV reverse transcriptase, 40 U RNasin, 2.5 U Taq polymerase added and incubated for twenty minutes at 43°C.

(3) After reverse transcription, 30 cycles of PCR are performed.

(4) Nested PCR can be performed on these samples by adding 5 μl of sample from the first round of PCR to a fresh reaction mix containing internal primers, fresh buffer, and dNTPs.

SPECIFIC PRIMERS FOR VIRUSES IN SLUDGE

Enteroviruses

5'   T g T C A C C A T A A g C A g C C   3'
5'   T C C g g C C C C T g A A T g C g g C T   3'

Internal sequence for seminested PCR

5′   C C C A A A g T A g T C g g T T C C g C   3′

PITFALLS AND PROBLEMS

(1) Sensitivity of detection: There are two problems with determining the sensitivity of virus detection by PCR. The first problem is that human enteric viruses are most likely indigenous to sewage sludge. Hence, unseeded sewage sludge, used as a negative control, may be PCR positive. To avoid this problem, samples should be autoclaved for thirty minutes and then subjected to PCR. If the unseeded, sewage-sludge sample becomes negative, sensitivity studies can be performed. Otherwise, the sample must be reautoclaved until the signal disappears.

The second problem could be termed "truth in sensitivity." For sensitivity studies, it is tempting to dilute the sewage sludge or sludge-amended soil with the virus, thus avoiding the problem of not being able to generate a negative control. The problem lies with the fact that as the inhibitors are diluted, sensitivity can increase by as much as 10–100-fold. In reality, the luxury of being able to dilute the sample may not be available due to low initial concentration of viruses and, thus, a false negative may be generated.

(2) Inhibitor removal: Both methods, the column chromatography and the antigen-capture methods, usually work for one sample treatment. However, for extremely dirty samples or high organic matter samples, inhibition may still be present. To verify this, seed a sample with from 1 to 10 PFU of virus and perform RT-PCR. If this sample fails to amplify after the first round, seminested PCR can be attempted. If even this method fails, the sample needs a second round of purification.

(3) Volume reduction issues: The method reported by Straub et al. [15] has an absolute sensitivity of 0.2 PFU, i.e., limited to the sample volume that can successfully be amplified using the current protocol. However, this translates to an equivalent of 200 PFU/g (dry wt.) for sludge-amended soil that can be successfully amplified. Attempts have been made to concentrate samples so that the equivalent number is equal to the absolute sensitivity. However, as samples become concentrated, two problems arise. Recovery efficiency decreased and PCR inhibition increases dramatically. Currently there is no solution to this problem.

(4) Amplification of non-cell, culture-infectious virus: Despite the problems outlined in item (3) above, Straub et al. [1] found that results were comparable to cell culture despite the small sample volume assayed by PCR. In fact, PCR detected viruses in two cases where cell

culture was negative. The enteroviruses, in particular, are notorious for producing defective viruses. For environmental samples, the ratio of cell-culture infectious versus noninfectious can be as great as $1{:}10^4$ [18]. This can become a problem for environmental samples in the sense that PCR results may all be positive, but cell culture results are all negative making the data difficult to interpret. In these cases other data may need to be obtained, such as the history of the samples.

APPLICATIONS IN ENVIRONMENTAL MICROBIOLOGY

These methods appear to have promising applications for monitoring potential viral pollution of land where sewage wastes are applied. Efficient methods are needed to concentrate samples and PCR methods need to be developed to determine if these viruses remain infectious to reconcile problems of detection via PCR and nondetection via cell culture.

REFERENCES

1 Straub, T. M., I. L. Pepper, and C. P. Gerba. 1994b. Detection of Naturally Occurring Enteroviruses and Hepatitis A Virus in Undigested and Anaerobically Digested Sludge Using the Polymerase Chain Reaction. *Can. J. Microbiol.* 40:884–888.

2 Pepper, I. L., K. L. Josephson, R. L. Bailey, M. D. Burr, and C. P. Gerba. 1993. Survival of Indicator Organisms in Sonoran Desert Soil Amended with Sewage Sludge. *J. Environ. Sci. Health.* A28:1287–1302.

3 Steffan, R. J. and R. M. Atlas. 1988. DNA Amplification to Enhance Detection of Genetically Engineered Bacteria in Environmental Samples. *Appl. Environ. Microbiol.* 54:2185–2191.

4 Josephson, K. L., S. D. Pillai, J. Way, C. P. Gerba, and I. L. Pepper. 1991. Fecal Coliforms in Soil Detected by Polymerase Chain Reaction and DNA-DNA Hybridizations. *Soil Sci. Soc. Am. J.* 55:1326–1332.

5 Pillai, S. D., K. L. Josephson, R. L. Bailey, C. P. Gerba, and I. L. Pepper. 1991. Rapid Method for Processing Soil Samples for Polymerase Chain Reaction Amplification of Specific Gene Sequences. *Appl. Environ. Microbiol.* 57:2283–2286.

6 Way, J. S., K. L. Josephson, S. D. Pillai, M. Abbaszadegan, C. P. Gerba, and I. L. Pepper. 1993. Specific Detection of *Salmonella* spp. by Multiplex Polymerase Chain Reaction. *Appl. Environ. Microbiol.* 59:1473–1479.

7 Neilson, J. W., K. L. Josephson, S. D. Pillai, and I. L. Pepper. 1992. Polymerase Chain Reaction and Gene Probe Detection of the 2,4-Dichlorophenoxyacetic Acid Degradation Plasmid, pJP4. *Appl. Environ. Microbiol.* 58:1271–1275.

8 Tsai, Y. L. and B. H. Olson. 1991. Rapid Method for Direct Extraction of DNA from Soil and Sediments. *Appl. Environ. Microbiol.* 57:1070–1074.

9 Holben, W. E., J. K. Jansson, B. K. Chelm, and J. M. Tiedje. 1988. DNA Probe Method for the Detection of Specific Microorganisms in the Soil Bacterial Community. *Appl. Environ. Microbiol.* 54:703–711.

10 Moran, M. A., V. L. Torsvik, T. Torsvik, and R. E. Hodson. 1993. Direct Extraction and Purification of rRNA for Ecological Studies. *Appl. Environ. Microbiol.* 59:915–918.

11 Ogram, A., G. S. Saylor, D. Gustin, and R. L. Lewis. 1988. DNA Adsorption to Soils and Sediments. *Environ. Sci. Technol.* 22:982–984.

12 Krawiec, S. and M. Riley. 1990. Organization of the Bacterial Chromosome. *Microbiol. Rev.* 54:502–539.

13 Ansari, S. A., S. R. Farrah, and G. R. Chaudry. 1992. Presence of Human Immunodeficiency Virus in Wastewater and Their Detection by Polymerase Chain Reaction. *Appl. Environ. Microbiol.* 58:3984–3990.

14 Kopecka, H., S. Dubrou, J. Prevot, J. Marechal, and J. M. Lopez-Pila. 1993. Detection of Naturally Occurring Enteroviruses in Waters by Reverse Transcription, Polymerase Chain Reaction, and Hybridization. *Appl. Environ. Microbiol.* 59:1213–1219.

15 Straub, T. M., I. L. Pepper, M. Abbaszadegan, and C. P. Gerba. 1994. A Method to Detect Enteroviruses in Sewage Sludge-Amended Soil Using the PCR. *Appl. Environ. Microbiol.* 60:1014–1017.

16 Graff, J., J. Ticehurst, and B. Flehmig. 1993. Detection of Hepatitis A Virus in Sewage Sludge by Antigen Capture Polymerase Chain Reaction. *Appl. Environ. Microbiol.* 59:3165–3170.

17 Soares, A. C., T. M. Straub, I. L. Pepper, and C. P. Gerba. 1994. Effect of Anaerobic Digestion on the Occurrence of Enteroviruses and *Giardia* Cysts in Sewage Sludge. *J. Environ. Sci. Hlth. Part A.* A29:1887–1897.

18 Sharp, D. G. 1965. Electron Microscopy and Viral Particle Function. In: *Transmission of Viruses by the Water Route.* Edited by G. Berg. Interscience Publishers, New York, pp. 193–217.

19 Tsai, Y. L., M. J. Park, and B. H. Olson. 1991. Rapid Method for Direct Extraction of mRNA from Seeded Soils. *Appl. Environ. Microbiol.* 57:765–768.

CHAPTER 5

Detection of Viruses in Water Samples by Nucleic Acid Amplification

MORTEZA ABBASZADEGAN[1]
RICARDO DeLEON[2]

INTRODUCTION

ENTERIC VIRUSES IN WATER

HUMAN enteric viruses are excreted in feces of infected individuals and may directly or indirectly contaminate water intended for drinking. Some of these viruses are excreted in high numbers, such as 10^6-10^8 per gram of feces of infected individuals. The enteric viruses include the enteroviruses, rotaviruses, Norwalk and Norwalk-like viruses, adenoviruses, reoviruses, and others. Surface and ground waters of the U.S. continue to be subjected to fecal contamination from a variety of sources, including sewage-treatment-plant effluent; on-site, septic, waste-treatment discharges, land runoff from urban, agricultural, and natural areas, and, possibly, leachates from sanitary landfills.

Evidence for fecal contamination of surface and ground waters is provided by the detection of enteric viruses in both surface and groundwater and the continued occurrence of outbreaks of waterborne disease. For example, between 1971 and 1985, 502 outbreaks of disease caused by waterborne contamination of the drinking water that involved 111,228 cases of illness were reported in the U.S. Of these, 49% were associated with groundwater sources and 51% were associated with surface-water sources [1,2]. Many of the reported outbreaks were due to enteric viruses (hepatitis

[1]Quality Control & Research Laboratory, American Water Works Service Company, Inc., 1115 South Illinois Street, Belleville, IL 67220, U.S.A.
[2]Water Quality Division, Metropolitan Water District of Southern California, 700 Moreno Avenue, La Verne, CA 91750-3399, U.S.A.

113

A virus, Norwalk virus, and rotaviruses). It is likely that many of the waterborne-disease outbreaks for which no etiological agent was identified (half of all reported outbreaks) were caused by viruses. This is the result of the failure to look for them and also, the limitations of current detection methodology.

ENTEROVIRUSES

The enteroviruses (poliovirus, coxsackie A and B viruses, echovirus) can cause a variety of illnesses ranging from gastroenteritis to myocarditis and aseptic meningitis [3]. Numerous studies have documented the presence of enteroviruses in raw and treated drinking water [4,5], wastewater [6], and sludge [7]. Enteroviruses in the environment pose a public-health risk because these viruses can be transmitted via the fecal-oral route through contaminated water [7] and because low numbers are able to initiate an infection in humans.

The enteroviruses are approximately 27 nm in diameter and have a positive polarity, single-stranded RNA genome of approximately 7,400 base pairs (bp). These viruses are relatively well known with regard to their replication cycle and genome organization. The full genome of several enteroviruses has been sequenced, thus permitting comparative nucleic acid and protein alignments [8].

HEPATITIS A VIRUS

Hepatitis A virus (HAV) is an important waterborne virus because of the severity of the disease it may cause in susceptible individuals. The HAV is the cause of acute infectious hepatitis and was the first enteric virus for which a waterborne outbreak was documented in the United States. This virus survives for more than four months at between 5°C and 25°C in water, wastewater, and sediments [9]. As with the enteroviruses, the full genome of various strains of HAV has been sequenced.

ROTAVIRUSES

Rotaviruses are a significant cause of acute diarrheal illness, especially in young children. Group A rotaviruses have been documented as causes of waterborne outbreaks in humans [10]. These viruses have a segmented genome consisting of eleven segments of double-stranded RNA. Segments designating subgroup and serotype specificity have been sequenced for several strains and serotypes.

CALICIVIRUSES

The calicivirus and the Small Round Gastroenteritis Virus (SRGV) groups have been implicated or suspected in several outbreaks of acute diarrheal illness. These groups are constituted of members such as Norwalk virus, Snow Mountain agent, Hawaii, Taunton, Parramatta and many other viruses that are yet unnamed [11]. Shared morphological and genomic characteristics of several of these viruses, such as having single-stranded RNA, single protein capsid, and genome organization similarities, have led to referring to these viruses as the Norwalk Group of viruses, with the Norwalk virus as its most typical member. The full genome of the Norwalk virus was first sequenced in 1990 [12], and partial sequences from various other Norwalk-like viruses have also been recently sequenced [13].

POLYMERASE CHAIN REACTION (PCR)

The PCR assays have been applied to the detection of enteroviruses in clinical [14,15] and environmental samples [16–18]. The PCR assays must be able to detect viruses after concentration from large volumes (from 100 to 1,500 l) of water [19]. This is usually accomplished by a filter-adsorption and elution method resulting in a concentrate containing viruses as well as organic and inorganic particulate material. Some of these compounds, such as humic substances, can interfere with the activity of the enzymes used in PCR.

CELL CULTURE METHODS

Conventional methodology for the detection of small numbers of enteric viruses from the environment relies on a few established cell lines. The BGM cell line is the most commonly used for the detection of enteroviruses in the environment. This cell line is preferred over others, including primary cells, because it provides high sensitivity to natural isolates of enteroviruses [20]. Its sensitivity can be further enhanced by pretreatment of the cells with enzymes or other substances [21]. Unfortunately, the use of other cell lines is required to detect other groups of enteric viruses [22]. This can greatly increase the cost and time of assay.

MATERIALS NEEDED

- virus stock (Virus strains can be obtained from the American Type Culture Collection, ATCC, Rockville, MD)

- primers for viruses
- PCR reagents and Taq polymerase
- phenol:chloroform:isoamyl alcohol
- Sephadex G-100
- Chelex 100
- water bath
- thermal cycler
- gel electrophoresis apparatus

ENTEROVIRUSES PRIMERS

Primers and probes for enteroviruses, hepatitis A virus, rotavirus and Norwalk viruses are listed in Table 1.

SAMPLE COLLECTION AND PROCESSING

SAMPLE COLLECTION

Water samples can be obtained by passing a large volume of water (100–1500 l) through a 1 MDS filter (CUNO Inc., Meriden, CT) at a flow rate of no more than 4 gallons per minute. The filter should be kept at $4°C$ and processed within forty-eight hours of the completion of sample collection.

FILTER ELUTION

The filter is eluted using an autoclaved solution of 1.5% beef extract (Becton Dickinson, Cockeysville, MD) 0.05 M glycine (U.S. Biochemical), pH 9.4. The solution is poured into the filter housing containing the 1 MDS filter. Allow fifteen minutes. The solution is then forced from the filter housing into a sterile, 2-l beaker using N_2. The eluate is then poured back into the filter housing and again forced out into the same beaker by using N_2. The pH of the solution is then lowered to 7.0–7.4 using 1 M HCl and stirring for fifteen minutes.

VIRUS FLOCCULATION AND RECONCENTRATION

The above solution is either stored at $-20°C$ or immediately adjusted to pH 3.5 and allowed to stir for fifteen minutes. The solution is then centrifuged for thirty minutes at 4,000 × g at $4°C$. The resulting pellet is resuspended in 0.15 M Na_2HPO_4 (pH 9.4) and transferred to a 50-ml centrifuge tube. The pH is adjusted to 7.2 and the volume brought to 15–30 ml

TABLE 1. Primers and Probes for Enteric Viruses.

Primers and Probes	Genomic Nucleotides	Sequence 5'-3'	Polarity	Amplicon size (bp)	Reference
Pan-enteroviruses					
5'-Nontranslated region					
Upstream primer	449–465	CCTCCggCCCCTgAATg	+	197	[18]
Probe	547–567	TACTTTgggTgTCCgTgTTTTC	+		
Downstream primer	627–644	ACCggATggCCAATCCAA	–		
5'Nontranslated region					
Upstream primer	445–465	TCCggCCCTgAATgCggCT		149	[16]
Probe	531–550	CCCAAAgTAgTCggTTCCgC			
Downstream primer	577–594	TgTCACCATAAgCAgCC			
Rotavirus Gene 4					
Upstream primer	676–695 (con1)	TTgCCACCAATTCAgAATAC	+	211	[27]
Probe	753–773 (AVP4-C)	AgAgAgCACAAgTTAATgAAg	+		
Downstream primer	887–868 (con2)	ATTTCggACCATTTATAACC	–		

(continued)

117

TABLE 1. (continued).

Primers and Probes	Genomic Nucleotides	Sequence 5'-3'	Polarity	Amplicon size (bp)	Reference
Hepatitis A virus (HM175) Capsid (VP1-VP3) region					
Upstream primer	2035–2054	CAgCACATCAgAAAggTgAg	+	192	[18]
Probe	2171–2192	TgCTCCTCTTTATCATgCTATg	+		
Downstream primer	2208–2226	CTCCAgAATCATCTCCAAC	−		
Norwalk virus 8FIIA Polymerase region				260	[26]
Upstream primer		CAAATTATgACAgAATCCTTC			
Probe		ATgTCATCAgggTCAAAgAgg			
Downstream primer		gAgAAATATgACATggATTgC			
Norwalklike viruses Polymerase region					
Upstream primer	4673–4692	gCACCATCTgAgATggATgT		209	[13]
Probe					
Downstream primer	4862–4881	gTTgACACAATCTCATCATC			

with 0.15 M Na_2HPO_4 (pH 7.2). This solution can be used for cell culture analysis and can be stored at $-70°C$ or can be used for PCR analysis.

CELL CULTURE ASSAY

The standard cell line to assay environmental samples for enteroviruses is the Buffalo Green Monkey Kidney (BGM) cell line. The BGM cells are grown to confluent monolayers in 25 cm^2 or 75 cm^2 plastic or glass flasks. Before exposure to the sample, the growth media are poured off and the cell monolayer is washed twice with Tris (Sigma Chemical Co., St. Louis, MO)-buffered saline solution. For each sample, a 1-ml or 3-ml volume of the final concentrate is inoculated into each of three replicate flasks. The 25 cm^2 or 75 cm^2 growth is incubated at 37°C for sixty minutes with rotational agitation every fifteen minutes to allow virus adsorption to the cells. Twenty milliliters of maintenance medium with 2% fetal-bovine serum and 1 ml of gentamycin (50 $\mu g/ml$) is added to each flask. The flasks are incubated at 37°C and examined daily for fourteen days for viral cytopathic effect (CPE). Any flask with suspected viral CPE is confirmed by passage into a fresh monolayer of BGM and observed for CPE.

All samples negative for CPE on the first passage can be passed a second time on BGM cells. All samples which exhibit CPE are confirmed by two additional passages on BGM cells. Each sample assay requires thirty to forty-five days to complete.

PROCEDURES

RT-PCR

The genome of most of the enteric viruses of interest is either single-stranded RNA (enteroviruses, HAV, Norwalk virus) or double-stranded RNA (rotaviruses). The RNA needs to be converted into DNA before amplification by the polymerase chain reaction. This step is accomplished by reverse transcriptase. During this first step reaction, the RNA is added to a reaction "cocktail" containing dNTPs, magnesium ions and either the downstream primer or random hexamers, reverse transcriptase, and RNase inhibitor in a RT-PCR compatible buffer. If the viral RNA is to be heat-released from the capsids just prior to reverse transcription, the enzyme, RNase inhibitor, the primer, and the magnesium are added after the tubes have cooled below 42°C.

The following procedures can be used for the reverse transcription and PCR. Procedures are based on single-tube reverse transcription (RT) and PCR. The RT reaction volume is increased to 30 μl to accommodate a larger sample volume. Ten microliters of sample concentrate plus 7.5 mM $MgCl_2$, $1 \times$ PCR amplification buffer ($10 \times$ buffer containing 500 mM KCl and 100 mM Tris-HCl [pH 8.3]) and deoxynucleotide triphosphates (dNTPs) at 200 μM each are added to a 0.5-ml tube. One hundred microliters of mineral oil is added on top of the mixture to prevent any sample loss during heating. The tube is heated at 99°C for five minutes to liberate genomic viral RNA from the viral protein coat, and then 50 U reverse transcriptase, 20 U RNase inhibitor, and 50 μM random hexamers (Promega) are added to the tube. Samples are placed in a DNA thermal cycler for the reverse transcriptase reaction with a temperature profile of 25°C for ten minutes, 42°C for forty-five to sixty minutes, and then 99°C for five minutes to completely denature the reverse transcriptase.

After the RT reaction, the PCR cocktail containing 2.5 U AmpliTaq DNA polymerase (Perkin-Elmer Cetus), $1 \times$ amplification buffer, 2 mM $MgCl_2$, primers at 0.5 μM each and double-distilled water is prepared and added underneath the mineral oil. The PCR amplification is performed by using a DNA thermal cycler with a temperature profile of 94°C for forty-five seconds, 55°C for thirty seconds and 72°C for forty-five seconds for a total of twenty-five to thirty cycles. The PCR products can be separated on a 1.6% agarose gel (FMC, Rockland, ME) and visualized using ethidium-bromide staining and/or southern hybridization (see Chapter 3).

LARGE-VOLUME POLYMERASE CHAIN REACTION

Most manufacturers of the enzymes needed for PCR describe reaction protocols in which the total reaction volume is from 30 to 100 μl. These are also the most commonly described reaction volumes in the scientific literature, although reactions of 10 μl or less are not uncommon. A trend has been observed to minimize the reaction volume to conserve reagents and to facilitate a greater number of reactions being run simultaneously. The drawback to this approach, when analyzing environmental samples, is the examination of smaller portions of a potentially dilute source.

The protocol discussed previously was followed by using 10 μl of a sample in 30 μl of reverse transcription reaction. This 10 μl represented 0.5 l of the original sample. As viral contamination may be at very low concentrations and still present health problems, it was desirable to maximize the sample size. This was accomplished by increasing the sample to 100 μl (representative of 10 l of sample/15 ml concentrate) without a 10-fold increase in the reaction size reagents. The amount of RNase inhibitor is only 3.3 times the amount used in the smaller reaction and the amount of reverse transcriptase is only 2 times the amount of the reaction, so a 10-fold increase in reaction volume is accomplished with about a 2.5-fold increase in cost. Additionally, the sensitivity of the reaction is greater and

TABLE 2. Reagents and Volumes for the Two RT-PCR Methods.

Reagents	Large-Volume PCR	Small-Volume PCR
Sample volume	10–200 μl	1–10 μl
Reverse transcriptase	2.5 μl[b]	1 μl[a]
RNasin	5 μl	1.5 μl
Taq polymerase	2 μl	1.0 μl
Equivalent volume assayed	5–10 l	0.5 l
Total volume of reaction	300 μl	100 μl

[a]AMV reverse transcriptase.
[b]Supercript II.

results more consistent [28]. See Table 2 for reagent volumes used for large and small-volume RT-PCR.

REVERSE TRANSCRIPTION

A total volume of from 10 μl to 200 μl of the Sephadex/Chelex-treated concentrate from above is added to 0 μl–190 μl of ddH$_2$O (to further dilute inhibitors) and 4 μl of random primers (500 μg/ml) (Promega Corp, Madison, WI). After mixing and brief centrifugation, the mixture is heated for four minutes at 96°C and then placed on ice. To this mixture add 30 μl of 10 × buffer (35 mM MgCL$_2$, 750 mM KCl, 100 mM Tris-HCl, pH 9.2), 30 μl of 0.1 M dithiothreitol, 8 μl of dNTP mixture (10 mM each dNTP) (Pharmacia), 5 μl RNasin (20–40 U/μl) (Promega Inc.), 2.5 μl Reverse Transcriptase (Superscript II, BRL Life Technologies, Gaithersburg, MD), and 10.5 μl ddH$_2$O to adjust the total volume to 290 μl. After mixing and brief centrifugation the mixture is incubated for ten minutes at 25°C, forty-five minutes at 42°C and five minutes at 99°C and is stored at 4°C until further use.

cDNA AMPLIFICATION

To the entire reaction above, add 2 μl of each of the upstream and downstream primers (75 μM each), 2 μl AmpliTaq DNA Polymerase (5 U/μl) (Perkin-Elmer, Norwalk, CT) and 4 μl ddH$_2$O. This mixture is vortexed, briefly centrifuged, and then overlaid with 100 μl mineral oil. The sample is placed in a thermal cycling machine and, after an initial denaturation of four minutes at 96°C, the sample is subjected to 35 cycles of 94°C for one minute and fifteen seconds, 55°C for one minute, and 72°C for one minute and fifteen seconds. These cycles are followed by seven minutes at 72°C and then stored at 4°C until the reaction is analyzed on a 1.6% agarose gel.

PCR OPTIMIZATION AND REMOVAL OF INHIBITORY SUBSTANCES

OPTIMIZATION OF RNA-PCR

The PCR technique is a sensitive reaction and when not optimized, nonspecific amplification, lower sensitivity, or inhibition of the reaction may occur. The reaction should be optimized for the reverse transcription (reverse transcriptase [RT], RNasin concentration) and amplification (number of PCR cycles, buffer concentration) protocols for the detection of small numbers of RNA viruses in water samples. The optimization of RT-PCR can be as follows:

(1) Amplification of low virus concentration in water concentrates:
 • units of reverse transcriptase, RNasin and Taq polymerase
 • magnesium and primer concentration (see Chapter 2)
(2) Temperature profiles:
 • template denaturation
 • annealing temperature
 • hot-start PCR
(3) Confirmation and controls:
 • southern hybridization
 • nested PCR
 • positive and negative controls (see Chapter 2)

To determine whether a water sample has high or low aquatic humic material, the specific UV absorbance (SUVA) can be measured. The SUVA is defined as the UV absorbance at 254 nm expressed as per meter of absorbance divided by the Dissolved Organic Carbon (DOC) concentration in mg/L. The SUVA above four indicates that the DOC of a water is composed largely of aquatic humics. The SUVA of under three indicates that the DOC is composed largely of nonhumic materials [23].

PHENOL-CHLOROFORM SAMPLE TREATMENT

A volume of 350 μl–750 μl of concentrate is added to an equal volume of phenol:chloroform:iso-amyl alcohol (Amresco Inc., Solon, OH). The mixture is vortexed for three minutes and then centrifuged for ten minutes at 12,000 $\times$ g at room temperature. Approximately 300–600 μl of the upper, aqueous layer is removed and added to 300–600 μl of chloroform, vortexed for one minute and centrifuged for ten minutes at 12,000 $\times$ g at room temperature. Approximately 500 μl of the upper, aqueous layer is removed for application onto the Sephadex column.

SEPHADEX G-100/CHELEX 100 TREATMENT

Sephadex columns are prepared as follows: 2.5 grams of Sephadex G-100 (Pharmacia Biotech, Piscataway, NJ) are equilibrated in 80 ml ddH$_2$O and allowed to sit at room temperature overnight before autoclaving at 121 °C for fifteen minutes. Five milliliters of the slurry is applied to a 5 cc syringe, into the bottom of which a small amount of sterile glass wool had been placed. The water is allowed to drip out of the column, leaving a bed of approximately 3 cc of Sephadex. The column is washed three times with 2 ml of ddH$_2$O. The 500 μl of sample from above is applied to the column and the 500 μl of effluent, designated Fraction 1, is discarded. Five hundred microliters of ddH$_2$O is then applied and Fraction 2 is discarded, as is Fraction 3, following a second application of 500 μl of ddH$_2$O. Another 500 μl of ddH$_2$O is applied and the effluent, Fraction 4, is collected in a 1.5-ml centrifuge tube. A small amount (about 50 μl vol.) of Chelex is added, the sample is vortexed, and then centrifuged for ten minutes before use in the RT-PCR phase of analysis.

THE PEG-PROCIPITATE-PEG METHOD

This method consists of concentrating viruses in the 1-l beef-extract eluent by polyethylene glycol (PEG) precipitation, followed by removal of PCR inhibitors by adsorption-elution of viruses to a commercial insoluble protein precipitate, ProCipitate, with a second PEG precipitation to reduce the volume of sample [24]. The main advantage of this method over those relying on RNA extractions is that intact virions are concentrated and purified. The final product can be used for RT-PCR amplification or for inoculation onto cell cultures if desired.

Supplement the 1-l eluent with 13 % PEG 8000 and 0.2 M NaCl. Mix the supplemented eluent at 4 °C for a minimum of two hours or overnight. Centrifuge at 7,000 × g at 4 °C for thirty minutes. Resuspend precipitate in 2–3 ml of sterile 5 mM KCl, 20 mM Tris-HCl (pH 8) buffer. Combine resuspended viruses with an equal volume of ProCipitate and mix for fifteen minutes to adsorb viruses to the insoluble material. Concentrate adsorbed viruses by centrifugation at 6000 × g at 10 °C for fifteen minutes. Resuspend and mix for one hour the virus-containing precipitate with at least 6 volumes of 0.1 M Tris-HCl at pH 9.0. Centrifuge at 6000 × g at 18 °C for fifteen minutes and retain the virus-containing supernatant. Concentrate viruses in the supernatant by a second PEG precipitation in 8 % PEG and 0.2 NaCl. Resuspend the precipitate in 0.2 ml of 5 mM KCl-20 mM Tris HCl pH 8. Viruses in the 0.2-ml product could be further purified by the methods outlined previously or in the next section, or a portion could be assayed by cell culture.

SEPHADEX G-200/ULTRAFILTRATION

Viruses concentrated by the PEG-ProCipitate-PEG method can be further purified prior to PCR by G-200 spin-columns and further concentrated if desired by ultrafiltration in Centricon microconcentrators. Prepare Sephadex G-200 columns (1-ml gel bed volume) as described by Maniatis et al. [24] except that silane treated glass wool is used to plug a 3 cc syringe and the centrifugal force is decreased to 400 × g to avoid crushing the soft, G-200 Sephadex beads (G-100 Sephadex could be also used, but G-200 removes PCR inhibitors more efficiently). Equilibrate the column with at least three successive washes with 5 mM KCl-20 mM Tris pH 8. Add 0.2 ml of concentrated sample and spin as in step 1. Collect sample in sterile 1.5-ml polypropylene tube. Purified viruses in the G-200 column-excluded volume could be assayed by PCR or further concentrated by ultrafiltration. A 1-ml Centricon microconcentrator (100,000 molecular weight cutoff) is pretreated with 0.1% Tween-80. The viruses in the column-excluded volume are added to the washed microconcentrator and centrifuged at 1000 × g until retentate volume is approximately 20–40 μl. The retentate can be used directly for PCR.

POSSIBLE PITFALLS

PCR cannot be performed on most of the concentrated water samples unless some of the organic material is removed prior to the PCR. The selection of the removal treatment must be based on applicability and high efficiency of the protocol.

The removal of potentially inhibiting material from the sample, either by chemical means, such as phenol/chloroform treatment, or by physical means, such as dilution and chromatographic separation, is critical to virus detection by RT-PCR. Although virus contamination in water may be at very low concentrations, RT-PCR techniques can potentially reveal the presence of viral RNA molecules. It has been shown, however, that some untreated environmental samples, even when seeded with artificially high virus concentrations, will mask detection of the viral RNA by this technique. A 100 μl environmental sample, which had been seeded with 10^2 PFU, was subjected to RT-PCR and failed to show any DNA amplification. The same sample diluted ten times and 100 times showed increasing amplification with each dilution.

While the detection of viral RNA does not indicate the level of contamination, the presence of viral RNA does indicate a source of viral contamination and, thus, the potential for health risk. The most sensitive method of detection would then seem to be the most desirable, even without the ability to confirm infectivity of the viruses present.

POSITIVE AND NEGATIVE CONTROLS FOR PCR ASSAYS

For each environmental sample, both a positive viral seeded sample, as well as a negative control sample, need to be run simultaneously with the environmental sample to allow reasonable interpretation of data. Negative amplification from a sample does not necessarily mean that no human enteroviruses are present in the sample. Controls must run through the same procedures as the samples to ensure that PCR inhibition is not occurring. The amplification of the sequence of enteroviruses in seeded samples suggests that the treatment protocol, and subsequently the RNA-PCR assay, are applicable for the detection of enteroviruses in the sample. For a more detailed explanation on the use of controls, refer to Chapter 2.

SUMMARY

PCR analysis of water samples for the detection of viruses is feasible provided the samples are first purified to remove some of the inhibitory substances from the water concentrate in order to facilitate the assay. The protocols for the inhibitor removal provided in this chapter will reduce the interfering materials for a successful PCR amplification. The large-volume protocol allows for sufficient removal or dilution of inhibitors so that over 95% of the samples are expected to amplify by PCR. The specificity and sensitivity of the reaction using the listed primers are sufficient to be utilized for testing of environmental samples for the detection of viruses. Seminested PCR and/or southern hybridization allow evaluation of samples with higher sensitivity and confirmation of the results.

REFERENCES

1 Craun, G. F. 1988. Surface water supplies and health. *J. Am. Water Works Association.* 80:40–52.
2 Craun, G. F . 1992. Waterborne disease outbreaks in the United States of America: causes and prevention. *World Health Statistics Quarterly.* Rapport Trimestriel de Statistiques Sanitaires Mondiales. 45(2–3):192–199.
3 Melnick, J. L. 1990. Enteroviruses: polioviruses, coxsackieviruses, echoviruses and newer enteroviruses, pp. 549–605. In: B. N. Fields (ed.), *Virology.* Raven Press, New York.
4 Keswick, B. H., C. P. Gerba, H. L. DuPont, and J. B. Rose. 1984. Detection of enteroviruses in treated drinking water. *Appl. Environ. Microbiol.* 47:1290–1294.
5 Keswick, B. H., C. P. Gerba, S. L. Secor, and I. Cech. 1982. Survival of enteric viruses and indicator bacteria in groundwater. *J. Environ. Sci. Health.* A17:903–912.
6 Payment, P. 1981. Isolation of viruses from drinking water at the Pont-Viau water treatment plant. *Can. J. Microbiol.* 27:417–420.

7 Craun, G. F. 1984. Health aspects of groundwater pollution, pp. 135-179. In: G. Bitton and C. P. Gerba (eds.), *Groundwater Pollution Microbiology.* John Wiley & Sons, Inc., New York.

8 Palmenberg, A. C., B. L. Semler and E. Ehrenfeld (eds). 1989. Sequence alignment of picornaviral capsid proteins. *Molecular Aspects of Picorna Virus.* ASM Press, Washington, D.C.

9 Sobsey, M. D., P. A. Shields, F. S. Hauchman, A. L. Davis, V. A. Rullman, and A. Bosch. 1988. Survival and persistence of hepatitis A virus in environmental samples, p. 121-124. In: A. J. Zuckerman (ed.) *Viral Hepatitis and Liver Disease.* Alan R. Liss, Inc., New York.

10 Gerba, C. P. and J. B. Rose. 1990. Viruses in source and drinking water. In: *Drinking Water Microbiology*, G. A. McFeters (ed.), Springer-Verlag, New York, pp. 380-395.

11 Kapikian, A.Z. and R. M. Chanock. 1990. Norwalk group of viruses. In: *Virology*, B. N. Fields et al. (eds.). Raven Press, Ltd. pp. 671-693.

12 Jiang, X. N., D. Y. Graham, K. N. Wang, and M. K. Estes. 1990. Norwalk virus genome cloning and characterization. *Science.* 250(4987):1580-1583.

13 Ando, T., S. S. Monroe, J. R. Gentsch, Q. I. Jin, D. C. Lewis, and R. I. Glass. 1995. Detection and differentiation of antigenically distinct small round-structured viruses (Norwalk-like viruses) by RT-PCR and southern hybridization. *J. Clin. Microbiol.* 33:64-71.

14 Hyypia, T., P. Auvinen, and M. Maaronen. 1989. Polymerase chain reaction for the human picornaviruses. *J. Gen. Virol.* 70:3261-3268.

15 Rotbart, H. A. 1990. Enzymatic RNA amplification of the enteroviruses. *J. Clin. Microbiol.* 28:438-442.

16 Abbaszadegan, M., M. S. Huber, C. P. Gerba, and I. L. Pepper. 1993. Detection of enteroviruses in groundwater with polymerase chain reaction. *Appl. Environ. Microbiol.* 59:1318-1324.

17 Pillai, S. D., K. L. Josephson, R. L. Baily, C. P. Gerba, and I. L. Pepper. 1991. Rapid method for processing soil samples for polymerase chain reaction amplification of specific gene sequences. *Appl. Environ. Microbiol.* 57:2285-2286.

18 DeLeon, R., C. Shieh, R. S. Baric, and M. D. Sobsey. 1990. Detection of enteroviruses and hepatitis A virus in environmental samples by gene probes and polymerase chain reaction. *Proc. 1990 AWWA WQTC,* San Diego, Calif. American Water Works Association, Denver, CO.

19 American Public Health Association (APHA), American Water Works Association, and Water Pollution Control Federation. 1989. *Standard Methods for the Examination of Water and Wastewater.* 17th Edition, Washington, D. C.

20 Dahling, D. R. and B. A. Wright. 1986. Optimization of the BGM cell line culture and viral assay procedures for monitoring viruses in the environment. *Appl. Environ. Microbiol.* 51:790-812.

21 Benton, W. H. and C. J. Hurst. 1986. Induction of cytopathogenicity in mammalian cell lines challenged with culturable enteric viruses and its enhancement by 5-iododeoxyuridine. *Appl. Environ. Microbiol.* 51:1036-1040.

22 Smith, E. M. and C. P. Gerba. 1982. Laboratory methods form the growth and detection of animal viruses. In: *Methods in Environmental Virology.* C. P. Gerba and S. M. Goyal, (eds.). pp. 15-48. Marcel Dekker, New York.

23 Edzwald, J. K. and J. E. Van Benschoten. 1990. Aluminum coagulation of natural

organic matter. *Proceedings of the 4th Gothenburg Symposium on Chemical Water and Wastewater Treatment,* Madrid, Spain.

24 Schwab, K. J., R. DeLeon, and M. D. Sobsey. 1995. Concentration and purification of beef extract mock eluates from water samples for the detection of enteroviruses, hepatitis A virus, and Norwalk virus by RT-PCR. *Appl. Environ. Microbiol.* 61:531–537.

25 Maniatis, T., E. F. Fritsch, and J. Sambrook. 1982. *Molecular Cloning: A Laboratory Manual.* Cold Spring Harbor Laboratory, Cold Spring Harbor, N.Y.

26 DeLeon, R., S. M. Matsui, R. S. Baric, J. E. Herrmann, N. R. Blacklow, H. B. Greenberg, and M. D. Sobsey. 1992. Detection of Norwalk virus in stool specimens by reverse transcriptase-polymerase chain reaction and non-radioactive oligoprobes. *J. Clin. Microbiol.* 30:3151–3157.

27 Gentsch, J. 1995. Personal Communication.

28 Abbaszadegan, M. and P. Stewart. 1994. Use of cell culture and RT-PCR for the detection of viruses in ground water. *Proc. 1994 SWWS/WQTC,* Denver, CO.

CHAPTER 6

Detection of *Giardia* Cysts and *Cryptosporidium* Oocysts in Water Samples by PCR

MORTEZA ABBASZADEGAN[1]

INTRODUCTION

THE enteric protozoa *Giardia* and *Cryptosporidium* are intestinal parasites that can cause gastroenteritis when they are ingested by humans. *Giardia* is the most common cause of parasitic infections in humans in the United States [1,2] and can cause lengthy bouts of diarrhea in infected individuals [3]. Numerous waterborne outbreaks of giardiasis and cryptosporidiosis have been documented [4–7]. Low numbers of *Giardia* cysts and *Cryptosporidium* oocysts are usually found in water supplies [2]. Cysts and oocysts are the infectious units of the microorganisms, and the development of approaches for the detection of these in water samples necessitates simple, efficient, and cost-effective methods. In addition, since an infection in humans can be initiated by as few as one to ten viable cysts [8], detection methods and viability assays need to be very sensitive. Currently, it is assumed that any *Giardia* and *Cryptosporidium* found in water is potentially infectious [9]. However, cysts or oocysts that are not viable do not pose a threat to human health. Therefore, it is important to be able to differentiate between viable and nonviable microorganisms.

The current methods for the detection of *Giardia* and *Cryptosporidium* in water rely primarily on microscopic observation of water concentrates using either phase-contrast microscopy [10] or an immunofluorescent technique [11], neither of which is able to differentiate between viable and non-

[1]Quality Control & Research Laboratory, American Water Works Service Company, Inc., 1115 South Illinois Street, Belleville, IL 67220, U.S.A.

viable cysts. The gene-probe technique [12] suffers from the same limitation.

Current methods to determine the viability of cysts include infectivity of animal models [13–15], in vitro excystation procedures [16,17], and the incorporation of vital dyes [18,19]. These methods are costly, time-consuming, and lack sensitivity because they require large numbers of cysts for the results to be statistically accurate. In addition, success differs from laboratory to laboratory and among *Giardia* strains [9]. Therefore, a more rapid and sensitive method for the detection of viable *Giardia* cysts in surface-water samples is needed for accurate and routine evaluation of water quality.

The polymerase chain reaction (PCR) [20,21] can be used to enzymatically amplify to detectable levels nucleic acid sequences that may be present in low copy numbers in water samples. The PCR has been used to determine cell viability and to detect the presence of mRNA, which has been correlated with the viability of *Legionella pneumophila* [22]. The PCR has also been used to determine *Giardia* viability by measuring the increase of total RNA after exposure of the cysts to excystation media for thirty minutes [23].

As a result of possible public-health threats, it is necessary to develop a rapid and sensitive assay for *Giardia* cysts and *Cryptosporidium* oocysts based on PCR-amplification of the mRNA of an inducible *Giardia* gene. The inducible gene used as an example in this chapter is an hsp70-like gene specific for *Giardia lamblia* that produces a heat-stress protein.

Many environmental stresses are known to induce the production of specific proteins called "stress proteins" that help protect an organism from damage until the stress is removed. Temperature change is an environmental stress that is known to activate a specific set of genes called the "heat-shock" genes, which are associated with newly synthesized mRNA [24,25]. Heat-shock mRNA appears in the cytoplasm of a viable cell within a few minutes of temperature elevation and is immediately translated with very high efficiency into a small number of highly conserved proteins, the heat-shock proteins (hsps) [26,27]. As long as cells are maintained at a high temperature, heat-shock proteins continue to be the primary products of protein synthesis. When cells are returned to a normal temperature, normal protein synthesis gradually resumes [24,27].

If a cell is not viable, it will not produce new heat-shock mRNA when it is exposed to elevated temperatures. The hsp70-like gene that is specific to *Giardia lamblia* has been identified and its promoter region determined [28]. By raising the temperature of the cysts, this hsp70-like gene can be induced to produce mRNA that codes for heat-shock proteins.

The purpose of this chapter is to describe practical approaches to the analysis of water samples for the detection of cysts or oocysts using PCR.

PROCEDURES

MATERIALS NEEDED FOR THE DETECTION OF *GIARDIA* CYSTS AND *CRYPTOSPORIDIUM* OOCYSTS

(1) Protozoan stock: *Giardia lamblia* cysts and *Cryptosporidium* oocysts
(2) Primers for *Giardia* cysts
(3) Primers for *Cryptosporidium* oocysts
(4) Liquid nitrogen
(5) Water bath
(6) Thermal cycler
(7) Reaction tubes, PCR reagents, Taq polymerase
(8) Gel electrophoresis apparatus

MATERIALS NEEDED FOR THE DETECTION OF VIABLE *GIARDIA* CYSTS

(1) mRNA isolation kit
(2) DNase and RNase enzymes
(3) Reverse transcriptase
(4) All the materials listed in the section above

Sample Collection

Water samples can be collected from a pressurized source or by using a portable, gasoline-driven water pump and filter housing in which are fitted ten-inch, yarn-wound polypropylene cartridge filters (U1A10U, Filterite Corp., Timonium, MD) with a nominal porosity of 1 μm. Flow rates need to be adjusted to one gallon per minute. A minimum of from 100 to 400 gallons of water needs to be filtered. After sample collection, the filter can be placed in a plastic bag, stored in a refrigerator, and processed within seventy-two hours.

Filter Elution

The filter is cut longitudinally, separated from the core, teased apart, and washed three times in 550 ml of elution medium. The washing is done on a shaker for ten minutes in a 1-gallon container. The washing procedure can be done using a Stomacher (Tekmar Co., Cincinnati, OH). The sample is concentrated and combined into a single pellet by centrifugation (2,500 × g for ten minutes). The final pellet is divided in half and a portion can be resuspended in 10% formalin for archiving.

Pellet Processing

Pellets are washed and resuspended in a solution containing 1% Tween 80 and 1% sodium dodecyl sulfate (SDS) solution and then homogenized (setting, 30) in a VirTis homogenizer (The VirTis Co., Inc., Gardiner, NY) for three minutes. One drop of antifoam (Sigma) is added to facilitate total sample recovery. Next, the sample is washed and resuspended in distilled water. Sample is sonicated (25 khz, Branson Ultrasonic Cleaner; Branson Sonic Co., Shelton, CT) and after sonication, 20 ml of each sample is underlayered with 30 ml of a density gradient solution Percoll-sucrose at a specific gravity of 1.10 [a mixture of 45 ml Percoll (Sigma, specific gravity 1.13) and 10 ml 2.5 M sucrose solutions; Sigma]. It is very important that the specific gravity be measured with a hydrometer (should be between 1.09 g/ml and 1.10 g/ml; if it is below 1.09, efficiency of recovery will be low). The gradient is then centrifuged at 1,040 × g for ten minutes. The entire supernatant from each sample is recovered, diluted 1:3 with 1% Tween 80, and centrifuged (1,200 × g for 10 min). The pellets are suspended in from 1 ml to 5 ml of 1% Tween 80. The pellet processing is an important part of the method. Recently, this step was modified. The modified procedure provides a better recovery of the microorganisms from water samples [29].

ALTERNATIVE METHODS FOR CONCENTRATION OF CYSTS AND OOCYSTS FROM WATER

Before analysis, parasites must first be concentrated from water to a volume that is easily assayed. The most commonly performed method of detection of protozoan microorganisms involves capture by filtration, filter washing, and concentration. In addition to this method, as described earlier, two other concentration methods can be used.

(1) The flocculation method reported by Vesey et al. [30] briefly states that 10–100 l of sample are concentrated by the addition of 1 M $NaHCO_3$ and 1 M $CaCl_2$ to the sample in order to give a final concentration of 0.01 M. The sample is then adjusted to pH 10.0 and allowed to stand until the floc settles (at least four hours). The supernatant is then siphoned off and the floc dissolved by the addition of 10% sulfamic acid. Cysts and oocysts are further reconcentrated by centrifugation.

(2) Samples can be concentrated by the membrane method [31] in which water is filtered through a 2 μm pore size, 293 mm diameter, flat polycarbonate membrane (Portics, Pleasanton, CA) in order to collect particles. The membrane is then washed and scraped to remove particles, which are further concentrated by centrifugation.

Treatment to Liberate Nucleic Acids from Cysts and Oocysts

Nucleic acids from *Giardia* cysts and *Cryptosporidium* oocysts can be extracted by freeze-thawing. A 10-μl aliquot of the stock can be serially diluted. Each dilution is then treated by freeze-thawing to liberate the nucleic acids from the microorganisms. Cysts or oocysts are frozen in liquid nitrogen for one minute and then thawed in a 65°C water bath for one minute. This procedure is repeated five times. The suspensions are then heated to 99°C for three minutes in a DNA thermal cycler (Perkin-Elmer Corp., Norwalk, CT) to denature proteins associated with the nucleic acids.

Primers for *Giardia*

The selection of the primers and the probe was based on computer alignment [32] of the *Giardia lamblia* sequences of genes for heat-shock protein 70, mRNA sequence (1,546 bp), rDNA fragment LSRNA, alpha-2 giardin gene and DNA for rRNA tandem repeat unit, and computer-assisted analysis of the primers. Five PCR primers, 17 to 22 base-pairs long, were derived from the *Giardia lamblia* gene for heat-shock protein 70 (hsp70). These five oligomeric strands were synthesized using an automated synthesizer. From the five oligomeric strands, a combination of three pairs of primers were selected and named hsp, hsp2, and hsp3 primers. At this point, the objective of having three pairs of primers for the detection of *Giardia* was to compare and evaluate the computer-assisted analysis of the primers with the laboratory results. This comparison enabled the selection of a better pair of PCR primers for the detection of *Giardia* cysts in water. The downstream primer of hsp, 5′-gTATCTgTgACCCgTCCgAg-3′, was synthesized "antisense" to the *Giardia* gene for hsp70, and the upstream primer, 5′-AgggCTCCggCATAACTTTCC-3′, was synthesized "sense" to the *Giardia* gene for hsp70, within a 163-base segment. The internal probe, 5′-gTgCAgCACAgAggCgCTgCTg-3′, was synthesized "antisense" to the *Giardia* gene for hsp70.

Primers for *Cryptosporidium*

Upstream primer 5′-AAgCTCgTAgTTggATTTCTg-3′; downstream primer 5′-TAAggTgCTgAAggAgTAAgg-3′; and the internal probe 5′-ggggATCgAAgACgATCAgATACCgTCgTAgTCTTAAC-3′ [33].

Reverse Transcription and Enzymatic Amplification

Procedures are based on single-tube reverse transcription (RT) and polymerase chain reactions (PCR). The RT reaction volume is increased to 30 μl to accommodate a larger sample volume. Ten microliters of sample plus 7.5 mM MgCl$_2$, 1 × PCR amplification buffer (10× buffer containing 500 mM KCl and 100 mM Tris-HCl [pH 8.3]) and deoxynucleoside triphosphates (dNTPs) at 200 μM each are added to a 0.5-ml tube. One-hundred microliters of mineral oil is added on top of the mixture to prevent any sample loss during heating. The tube is heated at 99°C for five minutes to denature any protein associated with nucleic acids, and then 50 U reverse transcriptase, 20 U RNase inhibitor and 50 μM random hexamers (GeneAmp RNA PCR kit, Perkin-Elmer, Norwalk, CT) are added to the tube. Samples are placed in a DNA thermal cycler (Perkin-Elmer) for the reverse transcriptase reaction with a temperature profile of 25°C for ten minutes, 42°C for forty-five to sixty minutes and 99°C for five minutes to completely denature the reverse transcriptase. After the RT reaction, the PCR cocktail containing 2.5 U AmpliTaq DNA polymerase (Perkin-Elmer), 1× amplification buffer, 2 mM MgCl$_2$, primers at 0.5 μM each and double-distilled water is prepared and added underneath the mineral oil. The PCR amplification is performed by using a DNA thermal cycler with a temperature profile of 94°C for forty-five seconds, 55°C for thirty seconds and 72°C for forty-five seconds for a total of thirty cycles. The PCR products can be separated on a 1.6% agarose gel (FMC, Rockland, ME) and visualized using ethidium-bromide staining (see Chapters 2 and 3).

Sensitivity of Reverse Transcription and Enzymatic Amplification (RT-PCR)

The sensitivity of RT-PCR procedure can be determined by the amplification of the 163 base pair region of heat-shock protein 70 of *Giardia lamblia*. A tenfold dilution of *Giardia* cysts is prepared and each dilution subjected to RT-PCR. In the laboratory, the procedure results in the detection of one single cyst.

Specificity of the Heat-Shock Protein Primers

The specificity of the primers within *G. lamblia, G. muris* and other microorganisms was examined. The microorganisms listed in Table 1 were screened for cross-reactivity.

RT-PCR FOR THE DETECTION OF VIABLE *GIARDIA* CYSTS (TABLE 2)

Treatments to Liberate Nucleic Acids from *Giardia* Cysts

Nucleic acids from *Giardia* cysts can be extracted by freeze-thawing. A 10-μl aliquot of the cyst stock is serially diluted and heated in a 42°C water

TABLE 1. Specificity of the HSP Primers for the Detection of *Giardia* Cysts.

Microorganisms	*Giardia*-Specific Amplification
Giardia lamblia (cyst)	+
Giardia muris (cyst)	+
Cryptosporidium parvum (oocyst)	−
Cryptosporidium muris (oocyst)	−
Ascaris lumbricoides (ova)	−
Ascaris suum (ova)	−
Hymenolepis nana (ova)	−
Trichomonas vaginalis (trophozoite)	−
Trichomonas foetus (trophozoite)	−
Hymenolepis nana (ova)	−
Clonorchis species (ova)	−
Hookworm (egg)	−
Trichoderma Species	−
Candida albicans	−
Rhizopus stolonifer	−
Aspergillus niger	−
Penicillium frequentans	−
Clostridium perfringens	−
Clostridium sporogenes	−
Salmonella typhimurium	−
Streptococcus faecalis	−
Bucillus subtilis	−
Klebsiella terrigena	−
Escherichia coli XL1-Blue	−
Escherichia coli ATCC 15597	−
Pseudomonas aeruginosa ATCC 27853	−
Adenovirus 41	−
Poliovirus type 1	−

The concentration of microorganisms was approximately 10^3 per reaction. The samples were washed and/or purified.

TABLE 2. Sensitivity of PCR Assay for the Detection of *Giardia* Cysts[a] in Water.

Amplification	Cyst Concentration[b]				
	10^3	10^2	10^1	10^0	10^{-1}
PCR	+	+	+	+	−
RNA-PCR[c]	+	+	+	+	−
mRNA-PCR[d]	+	+	+	−	−

[a]Cysts were purified by Percoll-sucrose density gradient.
[b]Cyst concentration was determined by immunofluorescence technique.
[c]This sample was treated by DNase before PCR assay.
[d]The mRNA was isolated by magnetic beads for PCR assay.

135

bath for twenty minutes to induce mRNA production for heat-shock proteins. Each dilution is then treated by freeze-thawing to liberate nucleic acids from the cysts. *Giardia* cysts are frozen in liquid nitrogen for one minute and then thawed in a 65°C water bath for one minute. The freeze-thaw procedure is repeated five times. The suspensions are then heated to 99°C for three minutes in a DNA thermal cycler (Perkin-Elmer Corp., Norwalk, CT) to denature proteins associated with the nucleic acids.

DNase AND RNase TREATMENTS OF LYSED CYSTS

The lysed cysts are treated with 0.1 μl RNase A (10 μg/μl) (Sigma) and incubated at 42° for fifteen minutes or with 0.5 μl of RNase-free DNase (1,000 U/ml, Promega) and incubated at 37°C for ten minutes. The RT-PCR is performed on each sample as described earlier.

mRNA EXTRACTION

Following the manufacturer's protocol, mRNA is recovered from the total nucleic acid using magnetic beads (Dynall, Oslo, Norway) with a poly dT tail covalently bound to each bead. As a control, the isolated mRNA can be treated separately with RNase A or DNase as described above in order to ensure that any amplification originates from mRNA and not DNA. After the treatment, RT-PCR is performed as described earlier. Results from one of these controls are shown in Table 3.

PCR AMPLIFICATION

A PCR protocol has been established to successfully amplify a 163 base-pair segment of the *Giardia* heat-shock protein 70 (hsp70) gene using the hsp primer set. The amplification has been optimized and is reproducible. The segment has been amplified from *Giardia* DNA coding for the gene, as well as from mRNA after induction of mRNA production, by

TABLE 3. RNase and DNase Treatment of Viable *Giardia lamblia* Cysts.

Enzymatic Treatment of Cysts' Nucleic Acids	PCR Results	
	mRNA-PCR	PCR
DNase I	+	−
RNase A	−	+
DNase I + RNase A	−	−

heating cysts at 42°C for fifteen minutes and isolation of *Giardia* mRNA using the mRNA isolation kit as described earlier. Amplification occurs in *Giardia* cysts that have been partially purified on a percoll sucrose density gradient. This purification method is very similar to the one used to isolate *Giardia* cysts from water concentrates in preparation for immunofluorescence detection of the cysts and oocysts.

Treatment of Water Samples for Removal of Inhibitory Substances

The PCR technique is a sensitive reaction and some physical (soil colloids) and/or chemical substances (organic or inorganic) may inhibit any amplification. The inhibitors should be removed and the reaction should be optimized for the reverse transcription and amplification protocols for the detection of small numbers of pathogens in water samples. Inhibitory substances from water concentrates have been removed in order to have a successful amplification of *Giardia* cysts and viruses in water samples [34] and sludge samples [35]. The RT-PCR and/or PCR cannot be performed on most of the concentrated water samples unless inhibitory factors are removed prior to the PCR. Amplification of *Giardia* cysts and *Cryptosporidium* oocysts that have been partially purified on a Percoll-sucrose density gradient, as described earlier, is successful. This purification method is very similar to the method used to isolate parasites from water concentrates in preparation for immunofluorescent detection of the cysts and oocysts in water samples. The purification steps seem to be sufficient for reducing inhibitory substances from water samples for PCR assay.

The removal of inhibitors is an important part of any successful PCR amplification of targeted sequences in environmental samples. This area requires a comprehensive research plan to evaluate removal of inhibitors in different types of water samples for PCR assay. The following can be used as an alternative to the density-gradient purification step for inhibitory removal and nucleic-acids extraction.

PHENOL-CHLOROFORM TREATMENT

Between 350 μl and 750 μl of sample concentrate are added to an equal volume of phenol:chloroform:isoamyl alcohol (Amresco Inc., Solon, OH). The mixture is vortexed for three minutes and then centrifuged for ten minutes at 12,000 × g at room temperature. Approximately 300 μl–600 μl of the upper, aqueous layer is removed and added to 300 μl–600 μl of chloroform, vortexed for one minute and centrifuged for ten minutes at 12,000 × g at room temperature. Approximately 500 μl of the upper, aqueous layer is removed for application onto a Sephadex column.

SEPHADEX G-100/CHELEX 100 TREATMENT

Sephadex columns are prepared as follows: 2.5 grams of Sephadex G-100 (Pharmacia Biotech, Piscataway, NJ) is equilibrated in 80 ml ddH$_2$O and allowed to sit at room temperature overnight before autoclaving at 121 °C for fifteen minutes. Five milliliters of the slurry is applied to a 5 cc syringe, with a small amount of sterile glass wool in the bottom (see Chapter 4). The water is allowed to drip out of the column, leaving a bed of approximately 3 cc of Sephadex. The column is washed three times with 1 ml of ddH$_2$O. The 500 μl of sample mentioned above is applied to the column and the 500 μl of effluent, designated Fraction 1, is discarded. Five hundred microliters of ddH$_2$O is then applied and Fraction 2 is discarded, as is Fraction 3, following a second application of 500 μl of ddH$_2$O. Another 500 μl of ddH$_2$O is applied and the effluent, Fraction 4, is collected in a 1.5-ml centrifuge tube. A small amount (about 50 μl) of Chelex is added, the sample is vortexed, and then centrifuged for ten minutes before use in the RT-PCR phase of analysis.

Specificity of the Detection Method

To evaluate the specificity of the *Giardia* heat-shock protein primers, *Giardia* species, as well as other parasites, bacteria, and viruses, were subjected to PCR and/or RT-PCR using primers flanking a sequence in the conserved hsp70 region of *Giardia lamblia* (Table 1). The 163 bp segment was not amplified in the other microorganisms. However, in some of the reactions, nonspecific amplification was observed. The nonspecific amplification was minimized by eliminating the first amplification cycle with annealing temperatures of 37 °C.

Sensitivity of the Detection Method

To determine the sensitivity of the reaction for the detection of *Giardia*, PCR was performed on *Giardia* cysts serially diluted in distilled water (Table 2) as well as in a surface-water concentrate sample. Both tests are necessary to determine the applicability of the system under optimal conditions as well as in an actual environmental sample. Similar sensitivity (one to five cysts) was achieved in seeded distilled water or surface-water concentrate after flotation as described earlier.

The PCR and RT-PCR procedure results in the detection of one cyst determined by ethidium-bromide staining. However, when mRNA is isolated using magnetic beads prior to RT-PCR procedure, the sensitivity drops to ten cysts (Table 2).

SUMMARY

The *Giardia* cysts and *Cryptosporidium* oocysts are the infectious units of the microorganisms, and the development of approaches for their detection in environmental samples necessitates simple, efficient, and cost-effective methods. The PCR is an attractive method to use for the monitoring of human pathogens in water samples because PCR is faster and simpler than the standard immunofluorescence.

The *Giardia* primers designed for this study were selected from the heat-shock protein 70 region of *Giardia lamblia* genome, which is a conserved region in the *Giardia* species. The RT-PCR amplification of the 163-bp region was diagnostic for *Giardia* cysts in water because results indicate that the primers are likely to be unique to *Giardia* cysts, and the 163-bp band was produced only when *Giardia* cysts were present in a sample. The specificity and sensitivity of the hsp primers selected for this study are sufficient to be used for routine environmental monitoring for the presence of *Giardia* cysts in water samples.

The sensitivity of PCR for the detection of cysts is comparable to the immunofluorescence techniques. However, the amplification occurs on *Giardia* cysts that have been partially purified on a percoll-sucrose density gradient, and this purification method is very similar to the method used to isolate *Giardia* cysts from water concentrates in preparation for immunofluorescent detection of the cysts in water samples. The RT-PCR and/or PCR cannot be performed on most of the concentrated water samples unless some of the organic material is removed prior to the PCR. The purification steps seem to be sufficient for reducing inhibitory substances from water samples for PCR assay.

For each environmental sample, both a positive *Giardia* cyst and *Cryptosporidium* oocyst seeded sample, as well as a negative control sample, need to be run simultaneously with the environmental sample in order to allow reasonable interpretation of data. Negative amplification from a sample does not necessarily mean that no cysts are present in the sample. Controls must be run through the same procedures as the samples to ensure that reverse transcription and/or PCR inhibition is not occurring (see Chapter 2).

A comparative study was performed to evaluate recovery of RNA from cysts using the total RNA isolation kit (Invitrogen) and mRNA isolation kit using magnetic beads (Invitrogen). The total isolation-kit procedure uses GuSCN, GHCl, ethidium bromide, and $CsCl_2$ density gradient. The recovery of RNA using the total isolation kit and, subsequently, the sensitivity of PCR was lower than when mRNA isolation kit (Invitrogen) was used.

One-cyst sensitivity in serially diluted samples has been repeatedly achieved. This sensitivity was achieved through RT-PCR amplification after the heat-shock protein gene was induced by heating. It seems induction increases the level of sensitivity for the detection of cysts in water samples.

It is suggested that for routine monitoring of cysts in water samples, RT-PCR be performed after heat induction of the samples. If any viable cyst is present in that sample, the level of sensitivity will be increased by mRNA production and, subsequently, an increase in the number of templates for PCR amplification. If mRNA is not produced by the heat induction in dead cysts, still DNA template of hsp70 gene will be amplified during PCR cycles for the present or absent analysis of cysts in water samples (Figure 1).

To assess the viability of cysts in an environmental sample, mRNA has to be purified by magnetic beads prior to RT-PCR amplification (Figure 2). The results of this study (Figure 3) suggest that only production of mRNA is associated with the viability of *Giardia* cysts in water samples. The mRNA could be isolated only from heat-induced *Giardia* cysts. Because of the protocol listed in Figures 1 and 2, one was able to determine if the

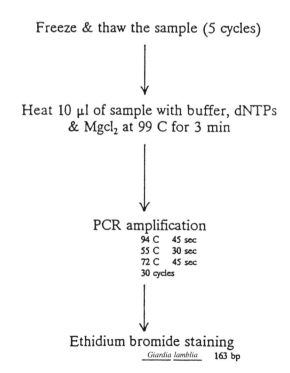

Freeze & thaw the sample (5 cycles)

Heat 10 μl of sample with buffer, dNTPs
& Mgcl₂ at 99 C for 3 min

PCR amplification
94 C   45 sec
55 C   30 sec
72 C   45 sec
30 cycles

Ethidium bromide staining
Giardia lamblia   163 bp

Figure 1 Protocol for the detection of *Giardia* cysts in water by PCR.

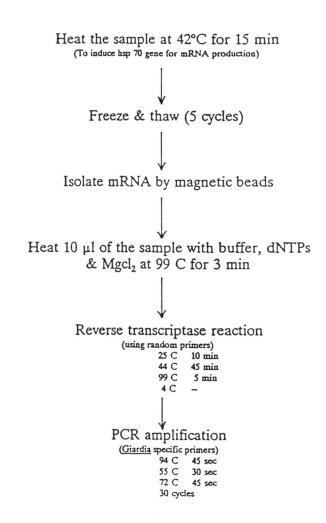

Heat the sample at 42°C for 15 min
(To induce hsp 70 gene for mRNA production)

Freeze & thaw (5 cycles)

Isolate mRNA by magnetic beads

Heat 10 µl of the sample with buffer, dNTPs
& Mgcl₂ at 99 C for 3 min

Reverse transcriptase reaction
(using random primers)
25 C	10 min
44 C	45 min
99 C	5 min
4 C	–

PCR amplification
(Giardia specific primers)
94 C	45 sec
55 C	30 sec
72 C	45 sec
30 cycles	

Ethidium bromide staining

Figure 2 Protocol for the detection of viable *Giardia* cysts in water by PCR.

M 1 2 3 4 5 6 7 8 9 10

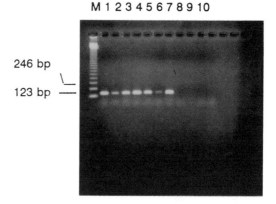

246 bp

123 bp

Figure 3 Ethidium bromide-stained agarose gel electrophoresis analysis of RNA-PCR and PCR amplification products using hsp primers. Lane M, marker; lane 1 and 2, PCR amplification of cysts; lanes 3–6, RT-PCR amplification of serial dilution of *Giardia lamblia* cysts; lane 7, purified mRNA-PCR amplification; lane 8, same sample as lane 7 but it was treated with RNase 1 before RNA-PCR assay; lane 9 and 10 negative controls.

amplification was originated from DNA or mRNA templates and, subsequently, differentiate dead or viable cysts in water samples.

In conclusion, the use of PCR for the detection of protozoan parasites in surface water is feasible provided the samples are first purified to remove some of the inhibitory factors from the water samples in order to facilitate PCR analysis. The specificity and sensitivity of the primers designed for this study are sufficient to be utilized for routine testing of environmental samples for the detection of *Giardia* cysts and/or *Cryptosporidium* oocysts.

REFERENCES

1 Craun, G. F. 1988. Surface water supplies and health. *J. American Water Works Association.* 80:40–52.

2 Craun, G. F. 1986. *Waterborne diseases in the United States.* CRC Press, Boca Raton, FL.

3 Wolfe, M. S. 1984. Symptomatology, diagnosis and treatment. In: *Giardia and Giardiasis.* S. L. Erlandsen and E. A. Meyer (ed.) Plenum Press, NY. pp. 147–161.

4 Moore, A. C., B. L. Herwaldt, G. F. Craun, R. L. Calderon, A. K. Highsmith, and D. D. Juranek. 1994. Waterborne disease in the United States, 1991 and 1992. *JAWWA.* 86:87–99.

5 Kent, G. P., J. R. Greenspan, J. L. Herndon, L. M. Mofenson, J. S. Harris, T. R. Eng, and H. A. Waskin. 1988. Epidemic giardiasis caused by a contaminated public water supply. *Am. J. Public Health.* 78:139–143.

6 Lippy, E. C. 1978. Tracing a giardiasis outbreak at Berlin, New Hampshire. *JAWWA.* 70:512-520.

7 Shaw, P. K., R. E. Brodsky, D. O. Lyman, B. T. Wood, C. P. Hibler, G. R. Healy, K. I. MacLeod, W. Stahl, and M. G. Schultz. 1977. A community-wide outbreak of giardiasis with evidence of transmission by municipal water supply. *Ann. Intern. Med.* 87:426-432.

8 Akin, E. W. and W. Jakubowski. 1986. Drinking water transmission of giardiasis in the United States. *Water Sci. Tech.* 18:219-226.

9 Hibler, C. P. 1988. Analysis of municipal water samples for *Giardia* cysts. In: *Advances in Giardia Research.* P. M. Wallis and B. R. Hammond (eds.), Univ. Calgary Press. pp. 237-245.

10 American Public Health Association (APHA), American Water Works Association, and Water Pollution Control Federation. 1985. *Standard Methods for the Examination of Water and Wastewater.* 16th Edition, Washington, D.C.

11 Sauch, J. F. 1985. Use of immunofluorescence and phase-contrast microscopy for detection and identification of *Giardia* cysts in water samples. *Appl. Environ. Microbiol.* 50:1434-1438.

12 Abbaszadegan, M., C. P. Gerba, and J. B. Rose. 1991. Detection of *Giardia* cysts with a cDNA probe and applications to water samples. *Appl. Environ. Microbiol.* 57:927-931.

13 Belosevic, M., G. M. Faubert, J. D. MacLean, C. Lau, and N. A. Croll. 1983. *Giardia lamblia* infections in Mongolian gerbils: an animal model. *J. Inf. Dis.* 147:222-226.

14 Labatiuk, C. W., F. W. Schaefer III, G. R. Finch, and M. Belosevic. 1991. Comparison of animal infectivity, excystation and fluorogenic dye as measures of *Giardia muris* inactivation by ozone. *Appl. Environ. Microbiol.* 57:3187-3192.

15 Roberts-Thomson, I. C., D. P. Stevens, A. A. F. Mahmoud, and K. S. Warren. 1976. Giardiasis in the mouse: An animal model. *Gastroenterol.* 71:57-61.

16 Bingham, A. K. and E. A. Meyer. 1979. *Giardia* excystation can be induced in vitro in acidic solutions. *Nature.* 277:301-302.

17 Boucher, S. E. M. and F. D. Gillin. 1990. Excystation of in vitro-derived *Giardia lamblia* cysts. *Infect. Immun.* 58:3516-3522.

18 Schupp, D. J. and S. L. Erlandsen. 1987. A new method to determine *Giardia* cyst viability: correlation of fluorescein diacetate and propidium iodide staining with animal infectivity. *Appl. Environ. Microbiol.* 53:704-709.

19 Smith, A. L. and H. V. Smith. 1989. A comparison of fluorescein diacetate and propidium iodide staining and *Giardia intestinalis* cyst viability. *Parasitology.* 99:329-331.

20 Mullis, K. B. and F. A. Faloona. 1987. Specific synthesis of DNA in vitro via a polymerase-catalyzed chain reaction. *Methods Enzymol.* 155:335-351.

21 Saiki, R. K., D. H. Gelfand, S. Stoffel, S. J. Scharf, R. Higuchi, G. T. Horn, K. B. Mullis, and H. A. Erlich. 1988. Primer-directed enzymatic amplification of DNA with a thermostable DNA polymerase. *Science.* 239:487-494.

22 Bej, A. K., M. H. Mahbubani, and R. M. Atlas. 1991. Detection of viable *Legionella* pneumophila in water by polymerase chain reaction and gene probe methods. *Appl. Environ. Microbiol.* 57:597-600.

23 Mahbubani, M. H., A. K. Bej, M. Perlin, F. W. Schaefer III, W. Jakubowski, and R. M. Atlas. 1991. Detection of *Giardia* cysts by using the polymerase chain reac-

tion and distinguishing live from dead cysts. *Appl. Environ. Microbiol.* 57:597–600.

24 Lindquist, S. 1986. The heat shock response. *Ann. Rev. Biochem.* 55:1151–1191.

25 Lindquist, S. 1980. Translational efficiency of heat-induced messages in *Drosophila melanogaster* cells. *J. Mol. Biol.* 137:151–158.

26 DiDomenico, B. J., G. E. Bugaisky, and S. Lindquist. 1982. The heat shock response is self-regulated at both the transcriptional and post-transcriptional levels. *Cell.* 31:593–603.

27 Lindley, T. A., P. R. Chakrabarty, and T. D. Edlind. 1988. Heat shock and stress response in *Giardia lamblia*. *Mol. Biochem. Parasitol.* 28:135–144.

28 Aggarwal, A., P. Romans, V. F. de la Cruz, and T. E. Nash. 1988. Conserved sequences of the HSP gene family in *Giardia lamblia*. In: *Advances in* Giardia *Research*. P. M. Wallis and B. R. Hammond (eds.) Univ. of Calgary Press, pp. 173–175.

29 LeChevallier, M. W., W. D. Norton, J. E. Siegel, and M. Abbaszadegan. 1994. Evaluation of the immunofluorescence procedure for detection of *Giardia* cysts and *Cryptosporidium* oocysts in water. *Appl. Environ. Microbiol.* 61:690–697.

30 Vesey, G., J. S. Slade, M. Byrne, K. Sheppard, and C. R. Fricker. 1993. A new method for the concentration of *Cryptosporidium* oocysts from water. *J. Appl. Bact.* 75:83–86.

31 Hansen, J. S. and J. E. Ongerth. 1991. Effects of time and watershed characteristics on the concentration of *Cryptosporidium* oocysts in river water. *Appl. Environ. Microbiol.* 10:2790–2795.

32 Higgins, D. G. and P. M. Sharp. 1988. Clustal: a package for performing multiple sequence alignments on a microcomputer. *Gene.* 73:237–244.

33 Pieniazek, N. 1995. Centers for Disease Control, Atlanta, GA (Personal Communication).

34 Abbaszadegan, M., M. S. Huber, I. L. Pepper, and C. P. Gerba. 1993. Detection of enteroviruses in groundwater with polymerase chain reaction. *Appl. Environ. Microbiol.* 59:1318–1324.

35 Straub, T. M., I. L. Pepper, M. Abbaszadegan, and C. P. Gerba. 1994. A method to detect enteroviruses in sewage sludge-amended soil using the PCR. *Appl. Environ. Microbiol.* 60:1014–1017.

CHAPTER 7

PCR Detection of Airborne Microorganisms

MARK P. BUTTNER[1]
ABDIEL J. ALVAREZ[1,3]
LINDA D. STETZENBACH[1]
GARY A. TORANZOS[2]

INTRODUCTION

THE term *bioaerosol* refers to an airborne suspension of microorganisms, microbial cell fragments, and by-products of metabolism [1]. Airborne microorganisms include bacterial cells, fungal spores or hyphae, viruses, algal cells or protozoan cysts. These organisms are not indigenous to the aerosphere but are being transported from one location to another through the air. Bioaerosols occur both indoors and outdoors and may be generated naturally or as a result of human activity [2,3]. Wave action aerosolizes microorganisms from surface and marine waters; and raindrop impaction and wind release organisms from plants and soil. Agricultural practices, such as tilling and harvesting, composting, and spray irrigation are examples of human activity resulting in the outdoor aerosolization of microorganisms. Indoors, microorganisms may be aerosolized from air-conditioning systems, food preparation, building materials and furnishings, as well as by occupants.

The composition and concentration of microorganisms in bioaerosols has generated interest in diverse areas, such as agricultural and industrial settings, health-care facilities, indoor air environments, and military research. Much of this interest is focused on the potential adverse health

[1]Harry Reid Center for Environmental Studies, University of Nevada–Las Vegas, Box 454009, Las Vegas, NV 89154-4009, U.S.A.
[2]Department of Biology, P.O. Box 23360, The University of Puerto Rico, Rio Piedras, Puerto Rico, 00931-3360.
[3]Current address: Abbott Diagnostics, Inc., P.O. Box 278, Barceloneta, Puerto Rico, 00617.

effects on humans resulting from bioaerosol exposure, such as allergic and hypersensitivity responses or infectious diseases. For example, bioaerosol monitoring is often conducted during indoor environmental quality investigations in office buildings and homes where occupants have experienced adverse health effects, which may be caused by microbiological contamination in the building [4]. This problem has received considerable attention since the outbreak of Legionnaire's disease in 1976 [5], and more recently the increased reporting of tuberculosis [6,7]. In addition, bioaerosol sampling is conducted in industrial settings (pharmaceutical and food processing) and health-care facilities for compliance monitoring and infection-control practices.

The biotechnology revolution has increased the use of microorganisms in manufacturing processes, where microorganisms are produced on a large scale in fermentation reactions or are applied to crops to enhance productivity. Specific and sensitive monitoring methods for the presence of genetically modified microorganisms and microbial pest-control agents released into the environment are needed because these organisms may drift from the target plot to surrounding areas [8]. Monitoring is needed in these areas to address regulatory concerns related to worker exposure and environmental exposure to bioaerosols [9].

The application of the PCR technique for enhanced detection of target microorganisms in bioaerosols is discussed in this chapter. Traditional bioaerosol sampling and analysis methods are presented, along with some of the concerns and limitations of these techniques. Air sampling and processing protocols, which have been used to adapt the PCR method to air sample analysis, are also discussed. Current methods used are outlined in detail, and there is a discussion of special considerations involving the use of PCR amplification for monitoring bioaerosols.

TRADITIONAL SAMPLING METHODS

The objective of bioaerosol monitoring is to measure the numbers and types of airborne microorganisms. Sampling for bioaerosols necessitates the separation of biological particles from the airstream and collection of the microorganisms for subsequent analysis. The three collection methods traditionally used in bioaerosol sampling are impaction, impingement, and filtration [10]. While each of these sampling methods relies on inertial forces to remove airborne microorganisms from the airstream, the collection medium is different in each case. With impactor samplers, airborne particles are collected on an agar or adhesive-coated surface. Impinger samplers utilize a liquid as the collection medium, and the filtration method is used to collect particles on a fibrous or membranous filter material. There are several types of aerobiological samplers commercially available [11,12], and the efficiency of these samplers has been previously reviewed [10,13].

The selection of a bioaerosol sampling method depends on several factors, such as collection efficiency of the sampler, the analysis method to be used, the estimated bioaerosol concentrations, equipment costs, portability of the sampler, and the length of time for air sample collection [4,11]. Addressing these questions prior to sampling can reduce problems with sample analysis and interpretation of results.

There are numerous concerns with traditional sampling and analysis methods that have led to the development of methods to enhance monitoring for bioaerosols. Most sampling methods currently in use for airborne bacteria and fungi rely on the enumeration of culturable microorganisms in order to determine bioaerosol concentrations. Airborne microorganisms, however, are subjected to stresses not only as a result of being in the air but also during the process of sampling. Some of these stressful conditions include: desiccation, osmotic shock, and mechanical shear forces. Sampling stress, combined with stresses incurred as a result of aerosolization, can result in a loss of viability or reduced culturability [14,15].

In addition, some bacteria (e.g., *Legionella* spp. and *Mycobacterium* spp.) require specific growth media and can take days or weeks to culture, which in many cases allows the nontarget microorganisms to overgrow and mask the organism of interest. Many environmental organisms cannot be cultured in the laboratory using current methods [16]. Organisms that are not culturable under the specific growth conditions imposed in the laboratory remain undetected, yet may be capable of causing illness.

Other analysis methods involving staining and direct microscopic enumeration of total cells [17] or spores [18] are tedious and time-consuming, and identification of the microorganisms may not be possible. Therefore, primary concerns with traditional methods are (1) airborne microorganisms may go undetected, resulting in false negatives and/or underestimation of bioaerosol concentrations, and (2) increased analysis time and effort may be required to detect the organism of interest.

THE PCR IN BIOAEROSOL MONITORING

Polymerase chain reaction (PCR) amplification is the novel utilization of a chemical sensor method to measure and monitor microorganisms [19]. While this technique has been used successfully to enhance the detection of microorganisms in other matrices, such as food [20], water [21], and soil [22], the use of this method for monitoring airborne microorganisms is a recent concept [23]. Application of the PCR technique to bioaerosol sampling permits the amplification of a target DNA sequence, eliminating the requirement for culture. One advantage of applying this method is an increase in sensitivity over traditional culture methods [24]. Positive results using the PCR technique have been demonstrated for the detection of bacteria aerosolized in a greenhouse when culture counts were negative [23]. Another advantage of the PCR is the rapidity with which results are obtained. It is possible to obtain results the same day of sample collection using PCR, compared with days or weeks for culture methods.

Currently, cost and equipment limitations for PCR make it necessary to use small reaction volumes. Therefore, a concentration step is required for PCR to be effectively applied to air sample analysis. Two sample collection methods, liquid impingement and filtration, have recently been successfully utilized for air sample collection with subsequent PCR detection of the target microorganism [23,25]. These two sampling methods will be discussed in the following sections.

LIQUID IMPINGEMENT SAMPLING

Liquid impingement samplers collect airborne particles into a liquid collection medium, usually a dilute buffered solution. The buffer is then used as an inoculum for culture methods or examined microscopically to determine the total concentration of airborne microorganisms. Several types of impinger samplers are available, such as the AGI-30 (Ace Glass, Inc., Vineland, NJ) and Greenburg-Smith samplers (Wheaton, Millville, NJ), which are operated at air-flow rates of 12.5 and 28.3 liters per minute (lpm), respectively. The AGI-30 sampler (Figure 1) was designed with a curved inlet to simulate airflow through the nasal passage and collect respirable microorganisms [14]. The cutoff size, or particle diameter above which it is assumed that all particles are collected, was calculated to be 0.31 μm for the AGI-30 [10]. The AGI-30 has been widely used for the collection of bioaerosols and has been compared to other aerobiological samplers in sampler efficiency studies [11-13, 26].

Recently, a solid-phase PCR (SP-PCR) method for the detection of airborne microorganisms was reported [23] in which the AGI-30 sampler was used to collect cells of a genetically engineered *Escherichia coli* strain that were aerosolized in a greenhouse. The resulting sample buffer was assayed for the presence of the target organism using both culture methods and SP-PCR. Aliquots of the collection buffer were assayed for culturable counts by membrane filtration of the sample and incubation of the membrane on a nutrient agar medium. Another aliquot of the sample was processed for SP-PCR amplification (Figure 2).

The use of SP-PCR [27], in which amplification takes place directly on a membrane filter, allowed the collection of the sample into a liquid and

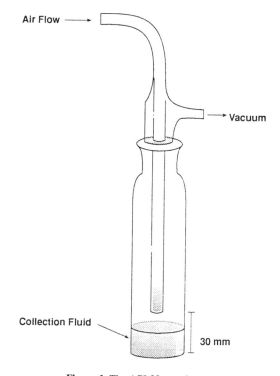

Figure 1 The AGI-30 sampler.

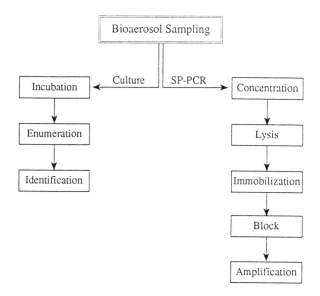

Figure 2 Outline of SP-PCR and culturable analysis for bioaerosol samples.

149

subsequent concentration onto a filter, which could be processed for PCR. With this procedure, results are obtained in fewer than eight hours, whereas culture methods require a minimum of forty-eight hours. The potential applicability of amplification of immobilized DNA for environmental monitoring of microorganisms in air samples has not been previously reported. The air sampling and PCR protocols used are described in the following sections.

Air Sampling Procedure

Materials

- sterilized all-glass impinger (AGI-30)
- vacuum pump and Tygon tubing with in-line flow meter [0–25 liters/minute (lpm) range]
- sterile tubes, pipettes, and pipette tips

Reagents

- 0.01 M potassium phosphate buffer (1.74 g KH_2PO_4 + 1.37 g K_2HPO_4 per liter of water) pH 7.5, autoclaved

Procedure

(1) Aseptically, pipet 20 ml of sterile 0.01 M potassium phosphate buffer, pH 7.5, into a sterilized AGI-30 (sampler inlet and outlet covered with aluminum foil).

(2) Support the sampler by clamping it to a ringstand in the desired sampling location.

(3) Remove the aluminum foil from the sampler outlet and attach Tygon tubing (or equivalent) with the calibrated in-line flow meter to the sampler and a vacuum pump.

(4) Set a timer to the desired sampling time. Remove aluminum foil from the inlet of the sampler.

(5) Activate the timer and vacuum pump simultaneously at the initiation of bioaerosol collection, and operate the sampler at an air-flow rate of 12.5 lpm for the appropriate sampling time.

(6) At the conclusion of the sampling period, replace the aluminum foil over the air sampler inlet, disconnect the tubing, and remove the sampler from the test area.

(7) Rinse the inlet with 5 ml of fresh sterile buffer to collect any particles remaining in the neck of the sampler. Disassemble the sampler and

aseptically pour the sample buffer into a 50-ml sterile tube. Place the sample buffer in a cooler with ice packs for transport to the laboratory.

(8) An aliquot of the buffer is assayed for culturable counts by membrane filtration using an appropriate nutrient medium for the target organism; another aliquot is used for the SP-PCR protocol.

Solid-Phase PCR Protocol

(1) Filter the sample through a 13-mm nylon membrane (MSI Magna, 0.2 μm pore size) placed into a sterile Millipore Swinnex-type or equivalent filter holder.

(2) Remove the membrane from the filter housing and place the side on which the cells are collected upwards on a piece of absorbent paper.

(3) To lyse the cells, the membranes are first placed faceup in an autoclavable dish lined with blotting paper saturated with 0.5 M NaOH/1.5 M NaCl and then steamed by placing the dish over boiling water for ten minutes.

(4) The membranes are removed from the steamed dish and sequentially treated for ten minutes at room temperature in dishes lined with blotting paper saturated with 0.5 M NaOH/1.5 M NaCl, 1 M Tris-HCl (pH 7.4)/1.5 M NaCl, and 0.3 M NaCl, respectively.

(5) After the last treatment, the membranes are again placed on absorbent paper to dry.

(6) Liberated DNA is then immobilized to the membrane by baking at 80°C for two hours or by an ultraviolet crosslinking step. If needed, at this point the membranes can be safely stored for future use by placing them in a Zip-Lock-type plastic bag or in a desiccator.

(7) To block nonspecific sites, the membrane is placed in a 1% BSA (BSA Fraction V; Sigma) solution 40°C in a shaking water bath, and incubated for a minimum of two hours.

(8) The membrane is removed from the blocking solution and placed on absorbent paper to dry. The dried membranes are placed in sterile 0.5-ml microcentrifuge tubes with the top side facing inward.

(9) The PCR mix can now be added. The PCR conditions are as follows: 100 mM Tris-HCl, pH 8.3, 500 mM KCl, 15 mM $MgCl$, 0.1% (w/v) gelatin, 200 μM dNTPs, 200 pmol each primer, 2.5 U Taq polymerase (Perkin-Elmer, Norwalk, CT), and ddH_2O to complete a total volume of 200 μl. Amplification conditions involve a denaturation temperature of 95°C, annealing at 52°C, and primer extension at 72°C, for a total of 30 cycles. The PCR results are detected after gel electrophoresis in a 1% agarose gel stained with ethidium bromide [28].

FILTRATION SAMPLING

The filtration method of bioaerosol sampling consists of impaction and attachment of the particles onto the filter medium [10]. Bioaerosols are most commonly sampled using membrane filters housed in disposable plastic cassettes. Filter membranes are available in a variety of materials, such as mixed cellulose ester, polycarbonate, and polyvinyl. The diameter of the membranes is usually 25 mm, 37 mm, or 47 mm, with pore sizes ranging from 0.01 μm to 10 μm depending on whether the sample will be used to detect viruses, bacteria, or larger microorganisms. Air sampling rates range from 1 lpm to 50 lpm. Filtration sampling results in a loss of viability of vegetative cells due to desiccation [13]; therefore, this sampling method is most often applied to determine total cell counts by staining and microscopic enumeration or culturable counts of fungal spores and endospore-forming bacteria that may be less sensitive to desiccation [12]. However, this method can be used in combination with the PCR technique if culturability is not a concern.

Recently, the filtration method of bioaerosol sampling has been applied for the detection of airborne varicella-zoster virus (VZV) in hospital rooms [25]. In that study, 37 mm mixed cellulose ester membranes with a pore size of 0.45 μm were used. Air was sampled at air-flow rates from 2.5 lpm to 9.4 lpm for from fifteen minutes to twenty-four hours. The membranes were then placed in deionized water and frozen at $-70\,^\circ$C. After thawing the samples, particles were eluted from the membranes by vortexing, and the sample was concentrated by centrifugation. After performing 2 freeze-thaw cycles, the sample was amplified by PCR. The VZV DNA was detected in 82% of samples from rooms housing patients with active varicella and 70% of rooms of patients with herpes zoster. Attempts to cultivate VZV from air filters were unsuccessful, probably due to desiccation of the virus during collection; thus it was not possible to determine what proportion of the VZV DNA detected was from infective viruses.

SUMMARY

To be useful, any method for bioaerosol monitoring must provide sufficient sensitivity and specificity to ensure detection of the target organism. Up to the present, the majority of existing data have been obtained using culture methods, which may underestimate the actual numbers of microorganisms present in aerosols. The use of the PCR technique for detection of airborne microorganisms is rapid, sensitive, and specific and can be considered as an alternative method for bioaerosol monitoring. As with many other techniques available to environmental laboratories, the decision to use nucleic acid amplification and/or conventional methods will likely be dictated by cost along with other factors, such as expertise. This is of special concern in developing countries that may lack capital, technology, and trained personnel to perform these kinds of studies. However, the cost per test will eventually be reduced by automation, increased sophistication, and market forces, leading to increased application in areas for which the cost/benefit ratio was previously limiting.

A promising application of the PCR technique in bioaerosol monitoring is the detection of airborne pathogens. The resurgence of antibiotic-resistant strains of *Mycobacterium tuberculosis* and the realization of the importance of aerosols as a means for its transmission make a very strong argument for the monitoring of areas in which this microorganism might be present. The detection of *Legionella* (discussed in Chapter 8) also becomes extremely important as a result of an increase in the number of immunocompromised members of the population. The monitoring of areas where highly infectious agents are cultured for vaccine production would benefit from the existence of methodology that is sensitive and rapid. In fact, PCR technology lends itself to automation.

The possible release of genetically modified microorganisms into natural environments has raised the issue of monitoring of these organisms from the standpoint of both human health and protection of the environment. It is expected that, in many instances, future release of these microorganisms into the environment will be as aerosols [1,29]. In the case of bacteria applied to field plots in aerosol form, the ability to specifically and sensitively predict aerial concentration and deposition patterns around the test site is needed. The validation of the SP-PCR technique, or other approaches, in field trials will help to generate data for risk assessment analysis concerning biotechnology agents and processes.

Although this chapter has focused on methods for the detection of bacterial cells and the protocols described result from the authors' experiences with these cells, the detection of viruses in aerosols is also an extremely important subject [30]. Although several viruses can be transmitted by air, the detection of airborne viruses has not been thoroughly studied. Because some enteric viruses may be transmitted as aerosols, more research is needed to understand the extent of infection due to airborne transport. In addition, little is known about the importance of air as a means for the transmission of rhinoviruses. Efficient collection of nanometer-sized particles is critical for virus bioaerosol monitoring, and the use of impingers and filtration methods for monitoring of viruses, as well as high-volume collection methods [31], need to be studied in detail. The PCR amplification lends itself to these types of studies.

Detection limits of < 10 CFU/m^3 are possible with the currently used samplers and culture analysis methods. Solid-phase PCR was found to be

TABLE 1. Primer Pairs for the Amplification of Selected Organisms.

Organism	Primer Sequences	Reference
Escherichia coli[a]	5′ CTgTTggCATCTTTTAgATTAAgTg 3′ 5′ gTTggTAgTTTggTAAAAgAATggT 3′	[23]
Escherichia coli	5′ TgTTACgTCCTgTAgAAAgCCC 3′ 5′ AAAACTgCCTggCACAgCAATT 3′	[33]
Varicella-Zoster virus	5′ TgACgATCTACTATTggAgAgAAC 3′ 5′ ATCCTgACCgTCCTCgCATACgTAg 3′	[25]
Legionella pneumophila	5′ gCATTggTgCCgATTTgg 3′ 5′ gCTTTgCCATCAAATCTTTCTgAA 3′	[37]
Mycobacterium tuberculosis[b]	5′ gAgATCgAgCTggAggATCC 3′ 5′ AgCTgCAgCCCAAAAggTgTT 3′	[38]

[a]Plasmid target.
[b]Clinical samples.

154

effective and more sensitive than the culture method for the detection of aerosolized bacterial cells in a greenhouse [23]. In addition to increasing the sensitivity of detection after amplification, hybridization with specific probes under high stringency conditions demonstrates the specificity of the amplification. Another benefit of the SP-PCR technique is that once the nucleic acids have been immobilized onto the filters, these filters can serve as the templates for several amplifications, unlike the liquid phase in which the templates can only be used once. The detection of several target microorganisms in the same sample is thus possible (e.g., multiplex PCR; [32]). This is important for all cases in which only a few copies of the target DNA are available or whenever the sample needs to be saved for later reamplification.

An alternative PCR method involving the lysis of cells using repeated freeze-thaw cycles [33] has been used for the detection of *E. coli* cells in air samples [34]. The sensitivity of the freeze-thaw and SP-PCR effects of environmental interferences in the protocol were determined [34].

The high sensitivity of the DNA amplification may have a negative aspect because this could result in a substantial increment of false positive results if appropriate precautions are not taken [35]. Ultraviolet irradiation provides a simple and efficient way to minimize contamination or false positives that often occur in laboratories performing routine PCR tests [36]. Meticulous laboratory techniques, along with judicious selection of controls [e.g., nontarget DNA samples (see Chapter 2)], will help minimize, if not eradicate, contamination problems. An intrinsic limitation of DNA amplification for bioaerosol monitoring is that viability of the airborne target organisms cannot be determined with current methods because the target sequences are amplified regardless of the nutritional or physiological status of the organism. At this time, monitoring studies need to include sampling methods that assess the viability aspects of bioaerosols. However, PCR can be used to improve the detection of nonviable airborne fungal spores that are not assessed with traditional, culture-based methods, but that can cause allergic responses [4].

It should be noted that the PCR can be used for the direct detection of any pathogenic or nonpathogenic microorganism in the air as long as unique, noncross-hybridizing sequences are available for the specific microorganisms to serve as primers for amplification (Table 1). Additional research to identify unique target sequences in microorganisms of health significance is needed.

REFERENCES

1 Stetzenbach, L. D. 1992. Airborne Microorganisms, pp. 53–66. In J. Lederberg (ed.), *Encyclopedia of Microbiology*. Academic Press, Harcourt Brace Jovanovich Publishers, San Diego, CA.

2 Buttner, M. P. and L. D. Stetzenbach. 1993. Monitoring of fungal spores in an experimental indoor environment to evaluate sampling methods and the effects of human activity on air sampling. *Appl. Environ. Microbiol.* 59:219-226.

3 Lighthart, B. and L. D. Stetzenbach. 1994. Distribution of microbial bioaerosols, pp. 68-98. In B. Lighthart and A. J. Mohr (eds.), *Atmospheric Microbial Aerosols, Theory and Applications*, Chapman and Hall, New York.

4 American Conference of Governmental Industrial Hygienists. 1989. Guidelines for the assessment of bioaerosols in the indoor environment. *American Conference of Governmental Industrial Hygienists*, Cincinnati, OH.

5 Morey, P. R. and J. C. Feeley. 1990. The landlord, tenant, and investigator: their needs, concerns, and viewpoints. In P. R. Morey, J. C. Feeley, Sr., and J. A. Otten (eds.), *Biological Contaminants in Indoor Environments*. American Society for Testing and Materials, Philadelphia, PA.

6 Center for Disease Control. 1994. Expanded tuberculosis surveillance and tuberculosis morbidity—United States, 1993. *MMWR.* 43:361-366.

7 Kaufmann, S. H. and J. D. van Embden. 1993. Tuberculosis: a neglected disease strikes back. *Trends in Microbiol.* 1:2-5.

8 Seidler, R. J. and S. Hern. 1988. *Special Report: Release of Ice-Minus Recombinant Bacteria.* U.S. EPA Environmental Research Laboratory, Corvallis, OR.

9 Sayre, P., J. Burckle, G. Macek, and G. LaVeck. 1994. Regulatory issues for bioaerosols, pp. 331-364. In B. Lighthart and A. J. Mohr (eds.), *Atmospheric Microbial Aerosols, Theory and Applications.* Chapman and Hall, New York.

10 Nevalainen, A., J. Pastuszka, F. Liebhaber, and K. Willeke. 1992. Performance of bioaerosol samplers: collection characteristics and sampler design considerations. *Atmospheric Environment.* 26A:531-540.

11 Chatigny, M. A. 1983. Sampling airborne microorganisms, pp. E1-E9. In P. J. Lioy (ed.), *Air Sampling Instruments for Evaluation of Atmospheric Contaminants.* American Conference of Governmental Industrial Hygienists, Cincinnati, OH.

12 Jensen, P. A., B. Lighthart, A. J. Mohr, and B. T. Shaffer. 1994. Instrumentation used with microbial bioaerosol, pp. 226-284. In B. Lighthart and A. J. Mohr (eds.), *Atmospheric Microbial Aerosols, Theory and Applications,* Chapman and Hall, New York.

13 Jensen, P. A., W. F. Todd, G. N. Davis, and P. V. Scarpino. 1992. Evaluation of eight bioaerosol samplers challenged with aerosols of free bacteria. *Am. Ind. Hyg. Assoc. J.* 53:660-667.

14 Cox, C. S. 1989. Airborne bacteria and viruses. *Sci. Prog. Oxford.* 73:469-500.

15 Marthi, B. 1994. Resuscitation of microbial bioaerosols. In B. Lighthart and A. J. Mohr (eds.), *Atmospheric Microbial Aerosols, Theory and Applications.* Chapman and Hall, New York.

16 Colwell, R. R., P. R. Brayton, D. J. Grimes, D. B. Roszak, S. A. Huq, and L. M. Palmer. 1985. Viable but non-culturable *Vibrio cholerae* and related pathogens in the environment: implications for release of genetically engineered microorganisms. *Bio/Technology.* 3:817-820.

17 Hobbie, J. E., R. J. Daley, and S. Jasper. 1977. Use of nucleopore filters for counting bacteria by fluorescence microscopy. *Appl. Environ. Microbiol.* 33:1225-1228.

18 Burge, H. P., J. R. Boise, J. A. Rutherford, and W. R. Solomon. 1977. Comparative recoveries of airborne fungus spores by viable and non-viable modes of volumetric collection. *Mycopathologia,* 61:27-33.

19 Saiki, R. K., D. H. Gelfand, S. Stoffel, S. J. Scharf, R. Higuchi, G. T. Horn, K. B. Mullis, and H. A. Erlich. 1988. Primer-directed enzymatic amplification of DNA with a thermostable DNA polymerase. *Science.* 239:487–494.

20 Wernars, K., E. Delfgou, P. S. Soentoro, and S. Notermans. 1991. Successful approach for detection of low numbers of enterotoxigenic *Escherichia coli* in minced meat by using the polymerase chain reaction. *Appl. Environ. Microbiol.* 57:1914–1919.

21 Bej, A. K., R. J. Steffan, J. DiCesare, L. Haff, and R. M. Atlas. 1990. Detection of coliform bacteria in water by polymerase chain reaction and gene probes. *Appl. Environ. Microbiol.* 56:307–314.

22 Tsai, Y.-L. and B. H. Olson. 1992. Detection of low numbers of bacterial cells in soils and sediments by polymerase chain reaction. *Appl. Environ. Microbiol.* 58:2292–2295.

23 Alvarez, A. J., M. P. Buttner, G. A. Toranzos, E. A Dvorsky, A. Toro, T. B. Heikes, L. E. Mertikas, and L. D. Stetzenbach. 1994. The use of solid-phase polymerase chain reaction for the enhanced detection of airborne microorganisms. *Appl. Environ. Microbiol.* 60:374–376.

24 Josephson, K. L., C. P. Gerba, and I. L. Pepper. 1993. Polymerase chain reaction detection of nonviable bacterial pathogens. *Appl. Environ. Microbiol.* 59:3513–3515.

25 Sawyer, M. H., C. J. Chamberlin, Y. N. Wu, N. Aintablian, and M. R. Wallace. 1994. Detection of varicella-zoster virus DNA in air samples from hospital rooms. *J. Infect. Dis.* 169:91–94.

26 Lembke, L. L., R. N. Kniseley, R. C. Van Nostrand and M. D. Hale. 1981. Precision of the all-glass impinger and the Andersen microbial impactor for air sampling in solid-waste handling facilities. *Appl. Environ. Microbiol.* 42:222–225.

27 Toranzos, G. A. and A. J. Alvarez. 1992. Solid-phase PCR for the detection of pathogens in water. *Can. J. Microbiol.* 38:365–369.

28 Sambrook, J., E. F. Fritsch, and T. Maniatis. 1989. *Molecular Cloning: A Laboratory Manual.* Cold Spring Harbor, NY: Cold Spring Harbor Laboratory Press, pp. 6.3–6.19.

29 Lindow, S. E., G. R. Knudsen, R. J. Seidler, M. V. Walter, V. W. Lambou, P. S. Amy, D. Schmedding, V. Prince, and S. Hern. 1988. Aerial dispersal and epiphytic survival of *Pseudomonas syringae* during a pretest for the release of genetically engineered strains into the environment. *Appl. Environ. Microbiol.* 54:1557–1563.

30 Spendlove, J. C. and K. F. Fannin. 1982. Methods of characterization of virus aerosols. In C. P. Gerba and S. M. Goyal (eds.). *Methods in Environmental Virology.* Marcel Dekker, New York.

31 Sorber, C. A. 1987. Recovering viruses from aerosols, pp. 53–66. In G. Berg (ed.), *Methods for Recovering Viruses from the Environment.* CRC Press, Inc., Boca Raton, FL.

32 Chamberlain, J. A., R. A. Gibbs, J. E. Ranier, P. N. Nguyen and C. T. Cashey. 1988. Deletion screening of the Duchenne muscular dystrophy locus via multiplex DNA amplification. *Nucleic Acids Res.* 16:11141–11156.

33 Bej, A. K., M. H. Mahbubani, J. L. DiCesare and R. M Atlas. 1991. PCR-gene probe detection of microorganisms using filter concentrated samples. *Appl. Environ. Microbiol.* 57:3529–3534.

34 Alvarez, A. J., M. P. Buttner and L. D. Stetzenbach. 1995. PCR for bioaerosol monitoring: sensitivity and environmental interference. *Appl. Environ. Microbiol.* 61:3639–3644.

35 Innis, M. A., D. H. Gelfand, J. J. Sninsky and T. J. White. 1990. *PCR Protocols: A Guide to Methods and Applications*. Academic Press, Inc., San Diego, CA.

36 Kwok, S. 1989. Procedures to minimize PCR-product carryover. *Amplifications*. 2:4–5.

37 Engleberg, N. C., C. Carter, D. R. Weber, N. P. Cianciotto and B. I. Eisenstein. 1989. DNA sequence of mip, a *Legionella pneumophila* gene associated with macrophage infectivity. *Infec. Immun.* 57:1263–1270.

38 Pierre, C., D. Lecossier, Y. Boussougant, D. Bocart, V. Joly, P. Yeni and A. J. Hance. 1991. Use of a reamplification protocol improves sensitivity of detection of *Mycobacterium tuberculosis* in clinical samples by amplification of DNA. *J. Clin. Microbiol.* 29:712–717.

CHAPTER 8

Detection of *Legionella* and *Legionella pneumophila* DNA in Environmental Water Samples Using the Polymerase Chain Reaction

TERESA PICONE[1]
THERESA YOUNG[1]
ELIZABETH FRICKER[2]

INTRODUCTION

IN 1976 a severe outbreak of pneumonia occurred among attendees of the American Legionnaires' Convention in Philadelphia, PA. Epidemiological investigation of the outbreak led to isolation and identification of a new gram negative bacterium, *Legionella pneumophila*. This bacterium was identified as the causative agent of the outbreak [1,2]. The bacterial source for this outbreak was attributed to a *Legionella*-contaminated cooling tower on the top of the hotel where the convention was held. In that outbreak alone, an estimated 182 people became ill and thirty-four died.

Prior to 1976, the organism, while not identified as *Legionella,* was already a causative agent of disease. In 1959 it was isolated from lung tissue of a diver in Key West, Florida [13]. In 1965 an outbreak of febrile disease was reported in a psychiatric hospital in Washington, D.C.; eighty-one cases were reported, twelve died [14,15]. Samples taken from these outbreaks were stored and later compared to those from the outbreak in Philadelphia [1,6]. These comparisons led to the conclusion that the causative agent in these outbreaks was *Legionella.*

Since 1976, *Legionella pneumophila* continues to be the causative agent of numerous outbreaks of disease [7–13] *Legionella pneumophila* and other species of *Legionella* are now associated with two disease states, Legionnaires' disease and Pontiac fever [1]. Legionnaires' disease

[1]Roche Molecular Systems, 1145 Atlantic Avenue, Alameda, CA 94501, U.S.A.
[2]Thames Water Utilities Ltd., Spencer House Laboratory, Manor Farm Road, Reading, Berkshire RG2 0JN, United Kingdom.

manifestations include pneumonia, fever, cough, and gastrointestinal symptoms. It usually occurs in individuals who are immunocompromised, and occurs more frequently in older males, particularly smokers and those who consume alcohol on a regular basis.

Pontiac fever is differentiated from Legionnaires' disease in that it is nonpneumonic, has a higher attack rate for healthy individuals, is self-limiting, and is not normally fatal. Pontiac fever is often not diagnosed since it may be asymptomatic or the symptoms may be so mild that they mimic a common cold.

Transmission of both diseases is most commonly associated with inhalation of *Legionella*-contaminated moist air from air-conditioning cooling towers [8] and other water sources, such as fountains [9], potable water [10], water misters [11], eyewash stations [12], and spas [13]. *Legionella* species are routinely found in all types of aqueous environments including lakes, rivers, oceans, and ponds [14–16, 67–68]. Isolation of *Legionella* from other sources, such as soil, have also been reported [17]. In addition, a recent outbreak of disease related to a cruise ship has been reported [18].

Over thirty species of *Legionella* have been identified [18–21], and many have been associated with human disease [17]. *Legionella pneumophila* is associated with 85% of the pneumonia due to *Legionella* [18]. This slow-growing organism requires cysteine and iron salts to grow, is resistant to chlorine (≤ 0.46 mg/L free chlorine), and viable at pH 2.2 for short periods of time (minutes). *Legionella* spp. have been demonstrated to grow within protozoa, such as *Acanthamoeba* [22–24]. Organisms within *Acanthamoeba prophaga* cysts have been shown to be even more resistant to chlorine (< 50 mg/L free chlorine) [25] than free-living cells.

Control of Legionnaires' disease and Pontiac fever is usually accomplished by treatment of the water and maintenance of the water reservoirs [26]. *Legionella's* resistance to chlorine makes decontamination of water reservoirs difficult. Recent outbreaks have indicated that current maintenance of some water reservoirs may be inadequate [7–13].

Culture techniques are traditionally used for detecting *Legionella* in environmental samples. Buffered charcoal yeast extract (BCYE) agar supplemented with cysteine, other various growth supplements, and antimicrobial agents are commonly used for routine isolation [27]. Culturing *Legionella* from environmental samples can be difficult and time-consuming because of the organism's characteristic slow-growth. Primary culture may require ten to fourteen days for colonies to form and the organism may not be detected. The sensitivity of culture methods is reported to be from 50% to 80% [13]. Researchers have reported that primary culture-negative samples were positive after sequential sampling and heat enrichment of the sample [28,29]. Viable *Legionella* are not always detectable by culture [30]. Other detection methods, such as Direct

Fluorescent Antibody (DFA) staining, DNA probe hybridization, and restriction enzyme digestion of DNA have been applied to environmental samples [31–37]. However, the specificity (due to cross reaction of the antibodies with non-*Legionella* organisms) and sensitivity (limited by the laborious microscopic examination of the sample) of these methods have not been sufficient for direct detection of *Legionella* in many environmental samples. Furthermore, unless viability markers are incorporated into the test, tests such as DFA will detect dead cells.

The polymerase chain reaction (PCR) amplification process has been shown to enhance the sensitivity of detection and the specific identification of target organisms including *Legionella* [38–45]. This process was used to develop a system for routine detection of 10–100 copies of *Legionella* DNA per ml by PCR of an original environmental water sample. Assuming that each *Legionella* cell contains one copy of *Legionella* DNA, this method would detect 10–100 *Legionella* cells per ml of original water sample. This system involves sample preparation, PCR amplification, and PCR product detection protocols. Detection of *Legionella* (requiring ten to fourteen days by culture) can be achieved in as little as five hours by PCR using the system described in this chapter.

SAMPLE PREPARATION

One of the keys to PCR amplification and detection is sample preparation. The sample preparation protocol should be easy to perform, generate a sample that can be used for PCR amplification, and recover the maximum amount of sample DNA. Many of the early successful sample preparation protocols were often clumsy, time-consuming, required special equipment, or used hazardous chemicals, such as phenol [43,46,47]. Even after processing the sample by these protocols, inhibitors of PCR amplification were not completely removed. Often, recovery of the sample nucleic acid was poor.

The sample preparation protocols described in this chapter incorporate a concentration step prior to extraction of the nucleic acid. This concentration step increases the sensitivity for detecting *Legionella* in the samples and also increases the statistical significance of the results. While a PCR amplification protocol could be developed with increased cycle numbers to achieve the same sensitivity, concentration also increases the amount of sample that will be tested, which increases the statistical significance of the results. Bacteria are concentrated by membrane filtration. Depending on the sample DNA preparation protocol chosen, from 10 ml to 100 ml of water sample is filtered. The filter membrane specified in this protocol was chosen because of its ability to trap *Legionella* and its inability to bind

DNA under the conditions tested. In this way the filter membrane separates DNA trapped within bacteria from free nucleic acid in the water sample. Both 0.2 μm and 0.45 μm filter pore membranes were tested. Recovery of *Legionella* by the two filters, as determined by PCR amplification, was comparable. Since the flow rate of the sample through the 0.45 μm filter was significantly faster, it was chosen.

After filtration, several sample DNA preparation protocols can be used. These protocols differ by the bacteria lysis method and the processing of the sample DNA after lysis. One protocol requires that the bacteria be washed off the surface of the filter in a suspension of Chelex resin followed by boiling. The Chelex protocol is very quick and easy to perform and works well for recovery of *Legionella* DNA from very clean water sources (such as potable waters). It does not work well for turbid waters where the required 100 ml sample volume can clog the filter. This protocol is also not very effective for reducing the inhibitory effects that environmental samples may have on PCR amplification. Studies with potable and cooling tower waters indicate that results can be generated for 75–96% of the environmental samples [48–52] depending upon the source and quality of the water. The remaining 4%–25% inhibit PCR amplification after processing by this protocol.

An alternative protocol involves lysis of *Legionella* in the presence of guanidine thiocyanate (GuSCN) [53] followed by alcohol precipitation. This protocol requires more time and care to perform but is very effective at reducing the inhibitory effect that environmental samples have on PCR amplification. Requiring only 10 ml of original water sample, the protocol achieves the same sensitivity as the Chelex protocol. Since it works well for a variety of sample, results could be generated for 96% of the potable and cooling tower waters tested with this protocol [54]. Addition of BSA to the amplification reaction and filtration of the processed sample can reduce the number of samples that inhibit PCR amplification after processing with GuSCN alcohol protocol to less than 1%. Details of these protocols are described in the methods section of this chapter.

When applying the Chelex or GuSCN methodologies to other applications, it is important to consider the source of the sample and the nucleic acid target. If the sample is from a relatively "pure" environment that does not contain a complex mixture of contaminating factors, due to its simplicity, the Chelex method would be preferred. If the sample is from a less pure source, the GuSCN method would be preferred. Additionally, while the protocols described in this chapter would function well for other DNA targets, they would not when the target sample nucleic acid is RNA. In this case, the GuSCN method would be preferred because this method would be more successful in removing proteins that would degrade RNA. For this protocol to be used for recovery of sample RNA, additional modifications of the GuSCN protocol would be required in order to preserve the integrity of the RNA. Such modifications should include: lowering the lysis temperature or modifying the method by which organisms in the sample are lysed, use of DEPC-treated water and RNAse-free containers, and immediate $-20°C$ storage of the samples.

PCR AMPLIFICATION

Several groups have utilized PCR for detection of *Legionella* [42–45]. Mahbubani et al. [43] selected two target genes for the detection of *Legionella*. The genus *Legionella* was identified using the 5S rRNA gene [55] and the species *L. pneumophila* was identified using the macrophage infectivity potentiator (*mip*) gene [56–58]. The primers developed by Mahbubani et al. did not, however, amplify both the 5S rRNA and *mip* gene targets at the same efficiency. Therefore, the same sensitivity for detecting the 5S rRNA and *mip* gene targets could not be obtained during coamplification (multiplex PCR) of both gene targets. That system also relied heavily on the detection probe for specific detection of *Legionella* 5S rRNA gene. The following changes were made to the Mahbubani et al. assay. These changes also illustrate issues to be addressed in the development of any robust amplification system.

The primers are optimized for more efficient specific multiplex PCR amplification of both targets. Primers with similar melting temperatures that annealed to regions with minimal secondary structure were selected. Smaller amplicons were used to increase the efficiency of amplification. Additional published and internally generated 5S rRNA [59,60] and *mip* [61] gene sequences were used for primer optimization. Where the sequences of the primer regions of different strains of *Legionella* differed, multiple primers were incorporated to PCR amplify the specific gene targets of these strains with the same efficiency.

Standard PCR reaction conditions are modified to optimize multiplex PCR amplification efficiency. A higher amount of Tris (50 mM) at a higher pH (8.9) is used in the PCR reaction. The amount of AmpliTaq® DNA polymerase is also increased. The increased amount of AmpliTaq DNA polymerase improves the efficiency of multiplex PCR amplification. Such a change was supported by reports that some causes of sample inhibition on PCR amplification can be overcome by using increased amounts of DNA polymerase [62].

Methods to prevent PCR carryover contamination are incorporated [63–65]. The PCR amplification reaction mix is optimized to incorporate dUTP instead of dTTP into the PCR amplified product. Incorporation of dUTP permits new PCR amplifications to be decontaminated from PCR product from a previous reaction. When the enzyme Uracil-N-glycosylase (UNG) is present, uracil will be hydrolysed from double-stranded DNA causing abasic residues. Upon heating, previously dU-containing DNA will break into small, unamplifiable fragments. Sample DNA that does not contain uracil will not be affected by UNG and will remain intact. The PCR products will be made from dT-containing DNA in the subsequent amplifications. Prior to amplification, the PCR samples are incubated at a temperature optimal for UNG activity followed by heating. In addition to helping to destroy the previously dU-containing DNA, the heating step also inactivates UNG. The PCR amplification is carried out at temperatures optimal for DNA polymerase activity where minimal UNG activity exists. After PCR amplification, the samples are stored at temperatures that prevent any residual UNG activity from affecting the newly amplified PCR product.

An internal positive control (IPC) was developed and incorporated into the PCR amplification reaction. This synthetic piece of DNA functions to prevent the reporting of failed PCR amplifications as false negatives and also permits the semiquantitation of the sample *Legionella* DNA. The IPC sequence, which is amplified by the *mip* primers, is presented in Figure 1. Because the IPC is part of the PCR amplification reaction, it will be amplified regardless of whether any *Legionella* genes are amplified. Absence of the 135 base-pair IPC PCR product band is an indicator of poor amplification due to either sample interference or incomplete preparation of the PCR reaction. The amount of IPC present in the PCR reaction mix has been titrated to permit semiquantitation of the sample DNA. Semiquantitation with the IPC is discussed further in the "Detection" section of this chapter.

The principles described in this section to optimize PCR amplification of *Legionella* DNA can be applied to developing other methodologies. Similar concentrations of buffer, magnesium, primer, dNTPs, and enzymes may be used as starting concentrations when optimizing other methodologies. The primers with similar melting temperatures (T_m), minimal secondary structure, and characteristics that are not prone to creating primer-dimer artifact with other primers in the master mix should be selected. These primers should have individual T_m values within a few degrees of each other and not be capable of base pairing with itself or other primers in the master mix. Depending on the application primer's T_m, the thermal cycler anneal and extend temperature may require modifications. Usually the thermal-cycler anneal and extend temperature is similar to the primer T_m but may differ by a few degrees. Internal positive controls can be designed and generated from a variety of sources. The internal positive control amplicon should be designed to be similar to the size of the target amplicon (see Chapter 2).

1 5'-GCATTGGTGC CGATTTGGGG AAGTTTGATG GAGATGAGGA GTTCTACGTG
negative control probe

51 GACCTGGAGA GGAAGGAGAC TGCCTGGCGG TGGCCTGAGT TCAGCAAATT
positive control probe

101 TGGAGGTTTT GTTCAGAAAG ATTTGATGGC AAAGC-3'

Figure 1 The internal positive control. The sequence of the IPC is shown with the locations of the positive and negative control probes. The negative control probe has a 1 base mismatch with the IPC sequence. The mismatch is underlined (the probe has a C instead of the correct G). The IPC primer regions are complementary to the *mip* primers.

DETECTION OF THE PCR PRODUCT

A nonradiactive, reverse dot-blot methodology [66] is used for the detection of the PCR product. When compared to other dot-blot protocols, this methodology reduces both the time and the effort required for detection. By using biotinylated 5S rRNA and *mip* primers, detection of the PCR-amplified product can be done without radioactivity. The PCR-amplified products generated by amplification with biotinylated primers specifically hybridize to immobilized probes. Use of short probes enhances their specificity. Under optimal conditions, stringent washing of the dot blots will allow only perfectly matched sequences to remain hybridized to the immobilized probe. The specifically bound biotinylated PCR products are detected by incubating with horseradish peroxidase conjugated to streptavidin (HRP-SA), washing and adding a substrate for horseradish peroxidase.

Specific probes and reagents are optimized for the detection of *Legionella* 5S rRNA and *L. pneumophila mip* gene PCR products. Two additional probes are immobilized on nylon membrane: a positive probe, whose sequence is perfectly complementary to the IPC sequence, and a negative probe, whose sequence differs from the IPC by one base. Presence of a positive-probe dot blot indicates that the PCR amplification and hybridization reactions were successful. Presence of the negative-probe dot blot or absence of a positive-probe dot blot may indicate that the stringent hybridization conditions were compromised. In this way, the IPC acts as an indicator of the reverse dot-blot hybridization conditions and prevents reporting of false positive or negative results due to poor, stringent, hybridization conditions. Additionally, the IPC has been titrated to create dot-blot intensities equal to that of 10^3 starting (prior to PCR amplification) copies of genomic *Legionella* DNA in the amplification reaction.

Dot-blot results from PCR amplified sample DNA of unknown *Legionella* concentration can be compared to the IPC dot-blot intensity. If the sample dot-blot intensities are greater than those of the IPC, then the amplification reaction contained greater than 10^3 starting copies; lighter than the IPC, less than 10^3 starting copies; equal to the IPC, 10^3 starting copies of *Legionella* DNA. These results would indicate that the sample contained $< 10^3$, 10^3 or $> 10^3$ cells per milliliter of original water sample (assuming that most *Legionella* contain one copy of DNA per cell and that 1 ml of unknown sample is incorporated into the amplification reaction using the protocols described in this chapter).

The reverse dot-blot methodology was chosen for detection of the PCR product because it permits nonradioactive detection of PCR products. When applying this method to other nucleic-acid targets, the length of the target nucleic acid's PCR product must be considered. During development of this protocol, shorter PCR products gave better sensitivity than longer ones. The capture probe length and hybridization conditions (salt and temperature) must also be considered if the probe is to be specific to the target PCR product. The protocol described in this chapter was designed to differentiate one base-pair mismatch between the probe and PCR product. In some applications, where the target nucleic acid is unique to the target organism (unlike the 5S rRNA gene), the stringency of the system may be relaxed. Additionally, the PCR products can be detected by a variety of other methods [42–45]. Agarose gel electrophoresis for detection of PCR products has also been used. The electrophoresis method described in this chapter is routinely used by our laboratory to detect from 0.1 kb to 1 kb PCR products. Detection of other longer PCR products may require lowering the percent of agarose in the gel. The sensitivity of gel electrophoresis, however, is approximately 10-fold lower than that of the reverse dot-blot method.

MATERIALS AND METHODS

CHELEX SAMPLE PREPARATION

One hundred ml of a water sample are filtered by vacuum through a 25 mm 0.45 μm Millipore Durapore® HVLP filter that has been autoclaved inside a Swinnex® Disc Filter Cartridge holder (Millipore Cat. No. SX00 02500). After filtration, the filter is aseptically removed and placed in an autoclaved 7 ml polypropylene tube (Bio-Rad Cat. No. 223-9800) containing 2 ml of freshly mixed DNA extraction reagent consisting of 20% (w/v) Chelex® 100 resin in 10 mM Tris-HCl, pH 8.0, 0.1 mM EDTA, 0.1% Sodium Azide. After vigorous vortexing for thirty seconds, the sample is placed in a beaker of boiling water for ten minutes followed by cooling and storage at 2°C–8°C for up to five days until required for PCR amplification. Twenty μl of the supernatant (the resin is not added) is used for PCR amplification.

GuSCN SAMPLE PREPARATION PROTOCOL

This protocol should be used instead of the Chelex Sample Preparation protocol when the sample is expected to contain high levels of humic acid, particulates, or other inhibitory substances. Ten ml of a water sample is filtered by vacuum filtration through a 25 mm 0.45 μm Millipore Durapore® HVLP filter that has been autoclaved inside Swinnex® Disc Filter Cartridge holders (Millipore Cat. No. SX00 02500). After filtration, the filter is aseptically removed and placed in an autoclaved, 2-ml-polypropylene microcentrifuge tube containing 0.5 ml of Lysis Reagent (containing the GuSCN) (PE Part No. N808-0134) and vortexed well to wash the bacteria from the surface of the filter. To facilitate lysis of the bacteria, the sample is incubated at 90°C for twenty minutes. After heating, the sample is cooled and centrifuged thirty seconds at 12,000 × g. Supernatant (400 μl) is removed and mixed with carrier reagent, 40 μg of RNA homopolymer poly A, to facilitate recovery of low concentrations of sample nucleic acid in a 1.5 ml microcentrifuge tube. Four hundred μl of isopropanol are added to the tube and mixed well. After a ten minute incubation, the sample is centrifuged ten minutes at 12,000 × g. The supernatant is removed from the tube and discarded. The remaining DNA pellet is washed once with 75% isopropanol and dissolved in 160 μl of autoclaved distilled water at 70°C for three minutes. The sample is ready for PCR amplification or may be stored at 2-8°C for up to five days, or at −20°C for up to two weeks until needed for PCR. The DNA extract (20 μl) is used for PCR amplification.

MODIFIED GuSCN SAMPLE PREPARATION PROTOCOL

Samples that inhibit PCR amplification after they are processed by the preceding GuSCN sample preparation protocol can be treated further. The processed sample is filtered through a Spin-X Centrifuge Filter Unit (0.22 μm, Costar Cat. No. 8160) by centrifugation for five to ten minutes at 12,000 × g to remove inhibiting particulates. Sample filtrate (100 μl) is mixed with 7 μl of 3% BSA solution (i.e., a 1:10 dilution of 30% BSA solution reagent from SERI Cat. No. A484). Twenty-one μl of this sample are used for PCR amplification. Alternatively, the Spin-X Centrifuge Filter Unit can be incorporated after the lysis of the bacteria and prior to the precipitation of sample DNA into the protocol described above and the BSA can be incorporated into the step that dissolves the DNA pellet in 160 μl of autoclaved distilled water.

PCR AMPLIFICATION

Each sample is added to 80 μl of PCR amplification reaction mix containing:

(1) 2 units of UNG (Perkin-Elmer Corporation (PE) part No. N808-0096)
(2) 4.5 units of AmpliTaq® DNA Polymerase (PE part No. N808-0060)
(3) 0.3 mM each of dUTP, dATP, dCTP and dGTP (PE part No. N808-0095)
(4) 50 mM Tris-HCl, pH 8.9, 50 mM KCl
(5) 15 mM $MgCl_2$

(6) 2.6 μM combined concentration of 7 oligonucleotide primers in two sets:

- three *mip* primers:
  PT69 (5'$_{biotin}$-gCATTggTgCCgATTTgg-3')
  PT70 (5'$_{biotin}$-gCTTTgCCATCAAATCTTTCTgAA-3')
  PT187 (5'$_{biotin}$- gTTTTgCCATCAAATCTTTTTgAA-3')
- four 5S rRNA gene primers:
  PT87 (5'-ggCgACTATAgCgATTTggAA-3')
  PT161 (5'-ggCgACTATAgCggTTTggAA-3')
  PT163 (5'$_{biotin}$-gCgATgACCTACTTTCgCATgA-3')
  PT165 (5'$_{biotin}$-gCgATgACCTACTTTCACATgA-3')

An Internal Positive Control DNA (IPC) sequence is shown in Figure 1.

The reactions are prepared in either Thin-Walled GeneAmp® Reaction Tubes (PE part No. N808–0611) for the Perkin-Elmer DNA thermal cycler 480 (TC480) or in MicroAmp™ Reaction Tubes (PE part No. N801–0612) for the Perkin-Elmer GeneAmp PCR System 9600 (TC9600). The reaction mixes in the Thin-Walled GeneAmp Reaction Tubes are overlaid with 75 μl of mineral oil (Sigma® Chemical Co. Cat. No. M3516). The samples are incubated for ten minutes at 45°C to facilitate UNG activity followed by ten minutes at 95°C. The samples are PCR amplified for 30 cycles of thirty seconds at 95°C and one minute at 63°C in the TC480, or for 30 cycles of fifteen seconds at 95°C and forty-five seconds at 63°C in the TC9600. After amplification, the samples are held at 72°C for a minimum of seven minutes but may be held overnight at 72°C prior to detection of PCR product. If longer sample storage is required, the samples may be stored at −20°C.

ELECTROPHORESIS

The PCR-amplified samples are analyzed by electrophoresis and/or reverse dot blot for the presence of amplified *Legionella* and IPC target DNA. For electrophoresis, 10 μl of amplified sample is mixed with 2 μl loading dye (40% Ficoll 400, 0.025% bromophenol blue) and loaded on an ethidium bromide containing 4% agarose gel (1% Nusieve/3% Seakem FMC Corporation). The samples are electrophoresed for two hours at 150 volts (see Chapters 2 and 3).

HYBRIDIZATION

The PCR product in the remaining 90 μl of the amplified sample is denatured by adding 72 μl of 0.4 N NaOH, 79 mM Na$_2$ EDTA·2H$_2$O, 0.005% thymol blue. Denatured sample (80 μl) is added to a hybridization tray well (PE part No. N808–0126) containing 2 ml of hybridization solution (0.6 M NaCl, 125 mM sodium phosphate, 5 mM EDTA, pH 6.8, 0.5% SDS) and probes are immobilized on a nylon membrane. The nylon membrane contains four different probe dots: (1) a genus-specific *Legionella* dot consisting of an equal blend of two probes that will hybridize to PCR amplified 5S rRNA gene product (PT125 5′-poly dT-gCgCCAATgA-TAgTgTg-3′ and PT127 5′-poly dT-gcgCCgATgATAgTgTg-3′); (2) a species-specific *L. pneumophila* dot that will hybridize to PCR amplified *mip* product (PT55 5′-poly dT-CATAgCgTCTTgCATgCCTTTAgCC-3′); (3) a negative IPC probe that differs by one base from a region of the amplified IPC (RH63 5′-poly dT-gATgAgCAgTTCTACgTgg-3′); (4) and a positive IPC probe that will hybridize to the PCR amplified IPC DNA (RH83 5′-poly dT-TgAgTTCAgCAAATTTggAg-3′). The contents of the tray well are incubated for twenty minutes in a shaking 55°C water bath. The liquid in the well is removed and 3 ml of enzyme conjugate HRP-SA (PE part No. N808–0092) in wash reagent (300 mM NaCl, 62.5 mM sodium phosphate, 2.5 mM EDTA, pH 6.8, 0.1% SDS) is added to the well. The tray is returned to the 55°C bath for twelve minutes. The nylon membranes are washed twice with 5 ml of wash reagent: once for ten minutes at 55°C and once for five minutes at room temperature. The membranes are subjected to a wash with 5 ml of 100 mM sodium citrate, pH 5.0 for five minutes at room temperature. After the sodium citrate wash, the membranes are incubated in 5 ml of color development solution (9.5 μg/mL 3,3′,5,5′-tetramethylbenzidine (TMB), 0.003% H_2O_2, 95 mM sodium citrate, pH 5.0) for thirty minutes at room temperature. After the color development solution, the membranes are washed well in deionized H_2O.

ENUMERATION OF *LEGIONELLA*

Legionella pneumophila SG1 cells were enumerated by culture on buffered charcoal yeast extract with alpha-ketoglutarate (BCYE-alpha) media. The cultures were incubated for up to five days at 37°C in 2.5% CO_2. The concentration (in colony forming units, CFU) was assigned.

Validated reagents used for sample preparation, PCR amplification, and detection of *Legionella* are now commercially available in kit form under the trade name EnviroAmp™ (Perkin-Elmer), the EnviroAmp *Legionella* Sample Preparation (Part No. N808–0088), PCR Amplification (Part No. N808–0089), and PCR Detection (Part No. N808–0090) Kits. These reagents were used in the studies described in this chapter.

RESULTS

The possible PCR amplification results of DNA extracted from *Legionella* species and *Legionella pneumophila* cells are presented in Figure 2.

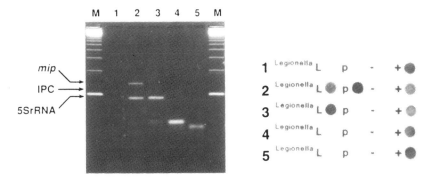

Figure 2 Agarose gel and hybridization results for amplified *Legionella* and *Legionella pneumophila* DNA. Lane 1 and Strip 1: 0 DNA added; #2: *L. pneumophila* DNA, 50 pg; #3: *L. dumoffii* DNA, 50 pg; #4: *Flavobacterium rigensi* DNA, 500 mg; #5: *Pseudomonas fluorescens* DNA, 500 mg; Lane M: 123 molecular weight marker.

The PCR products for the *Legionella* 5R rRNA gene, *Legionella pneumophila mip* gene and IPC DNA are 107, 167 and 135 base pairs in length, respectively. The reverse dot-blot strips contain the 5S rRNA gene probe immobilized next to the "L," the *mip* gene probe immobilized next to the "p," the negative IPC probe next to the "−," and the positive IPC probe next to the "+" symbols on the membrane. Presence of a 5S rRNA agarose gel PCR product band, or a blue dot next to the "L" on the membrane, is indicative of the presence of organisms of the genus *Legionella*. Presence of both 5S rRNA and *mip* PCR product bands, or blue dots next to both the "L" and "p," are indicative of the presence of *Legionella pneumophila*. The PCR-amplified IPC product band, or a blue dot next to the "+," should always be present. Absence of the IPC PCR product band or "+" dot indicates poor PCR amplification. Poor amplification can be due to the presence of inhibitors or due to incomplete preparation of the PCR amplification reaction. Presence of a dot next to the "−" on the membrane indicates that stringent hybridization conditions have been compromised.

The PCR amplification and reverse dot-blot detection have been optimized to obtain the best possible balance in band and dot intensities. The agarose gel band and reverse dot-blot intensity of the 5S rRNA PCR product can be slightly stronger than that of the *mip* PCR product at the same starting copy of genomic *Legionella pneumophila* DNA. This slight difference in intensity is because of the presence of multiple copies of 5S rRNA genes within one copy of genomic *Legionella* DNA. Slightly darker dot intensities for the *mip* ("p") PCR product dot can occur during hybridization as a result of a slightly warmer water bath. Significantly greater differences between the 5S rRNA and *mip* PCR product bands is indicative of a sample containing a mixture of non-*L. pneumophila* and *L. pneumophila* DNA. The reader should take this into consideration when trying to semiquantitate the results.

The results from PCR amplification reactions of samples containing DNA extracts of varying concentrations of *Legionella pneumophila* cells are presented in Figure 3. The concentration of *Legionella pneumophila* was determined by culture. The ethidium-stained PCR product bands and reverse dot-blot intensities increase with increasing concentrations of *Legionella pneumophila* cells. Because the IPC DNA is constant in the reaction mix, IPC PCR product bands and dot-blot intensities remain relatively constant and can be used to quantify the amount of sample *Legionella pneumophila* DNA. This is achieved by comparing the intensity of the IPC dot to that of the genus "L" and species "p" dots. The "L" and "p" dots darker than the IPC indicate that the 20 μl amplification sample contains greater than 1000 *Legionella pneumophila* cells; lighter dots indicate less than 1000 *Legionella pneumophila* cells are present.

During processing, an environmental sample is concentrated 50-fold. Therefore, the 20 μl sample used for PCR amplification represents 1 ml of original water sample. The protocol described in this chapter can readily detect as few as 10 to 100 colony-forming units (CFU) of *Legionella pneumophila* DNA per 20 μl PCR amplification sample or from 10 to 100 CFU/ml in the original water sample. While this protocol uses filtration to concentrate the sample, centrifugation may also be used. Additionally, even though each PCR amplification reaction only tests 1 ml of original water sample, due to the variable nature of the sample, better statistical sampling for the presence of *Legionella* in the original water sample can be achieved by testing larger volumes of the original water. This protocol was optimized for this level of sensitivity based on discussions with scientists who were concerned that a more sensitive PCR method may be difficult to interpret given the ubiquitous nature of *Legionella*. It has also

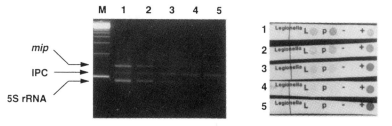

Figure 3 Detection of varying concentrations of *Legionella pneumophila*. Lane 1: 11,200 *L. pneumophila* colony forming units (CFU)/ml; Lane 2: 1120 *L. pneumophila* CFU/ml; Lane 3: 112 *L. pneumophilia* CFU/ml; Lane 4: 11 *L. pneumophila* CFU/ml; Lane 5: 0 *L. pneumophila* CFU/ml; Lane M: 123 bp molecular weight marker.

TABLE 1. Specificity for *Legionella*.

L. anisa	*L. birminghamensis*	*L. bozemanii*	*L. cherrii*
L. cincinnatiensis	*L. dumoffi*	*L. erythra*	*L. feelii*
L. gormanii	*L. gratiana*	*L. hackeliae*	*L. jamestowniensis*
L. jordanis	*L. longbeachae*	*L. maceachernii*	*L. micadadei*
L. moravica	*L. oakridgensis*	*L. parisiensis*	*L. quinlivanii*
L. sainthelensi	*L. spiritensis*	*L. tucsonensis*	*L. wadsworthii*
L. pneumophila (Serogroups 1 thru 15)			

These species of *Legionella* are known to be amplified and detected. With the exception of *L. hackeliae*, none of the nonpneumophila species of *Legionella* will produce a *mip* PCR product of a positive "p" dot. All species are amplified at the same efficiency except *L. spiritensis*, *L. quinlivanii*, and *L. sainthelensi*. *L. israelensis* is not detected with these primers and probes.

been suggested that cooling-tower waters containing 10^3 organisms per milliliter, or greater, represent levels of increased health risk [67]. Additional sensitivity can be obtained by increasing sample concentration, the amount of sample for PCR amplification or detection sample, or the number of PCR amplification cycles. Sensitivity, however, may be compromised by the quality and source of the water sample.

A summary of the results obtained for PCR amplification and detection of different *Legionella* and non-*Legionella* strains are presented in Tables 1 and 2. While other species of the genus *Legionella* have been reported to contain "*mip*-like" sequences [57], only one non-*L. pneumophila* strain, *L. hackeliae,* is amplified and detected by the *mip* primers and probes. Three *Legionella* strains (*L. spiritensis, L. quinlivanii,* and *L. sainthelensi*) are not efficiently PCR amplified by the 5S rRNA primers resulting in reduced sensitivity. *L. israelensis* cannot be detected by the 5S rRNA primers and probes described here. Similarity of *L. israelensis* 5S rRNA sequences to other non-*Legionella* organisms prevents the design of primers and probes that detect this species' 5S rRNA gene without compromising the specificity of this method for *Legionella*. Only one non-*Legionella* organism, *Alteromonas rubra,* has been shown to cross-react with the 5S rRNA primers and probes. In order to be detected, *Alteromonas rubra* must be present in the original water sample at concentrations greater that 10^5 CFU/ml.

A comparative study was performed at Thames Water Utilities, U.K., where the EnviroAmp™ Legionella Sample Preparation, PCR Amplification and Detection Kits' results were compared with culture results for eighty-seven samples of naturally contaminated water (Table 3). Prior to PCR amplification, the samples were processed using the GuSCN-sample preparation protocol. Any of the GuSCN-prepared samples that inhibited PCR amplification were treated with BSA, filtered, and retested at Roche. None of the samples inhibited PCR amplification after the additional BSA

TABLE 2. Non-Legionella Species Tested.

Acetobacter aceti
Acetobacter aurentatius
Acetobacter pasteurianus
Acinetobacter calcoaceticus
Acinetobacter iwoffi
Aeromonas hydrophila
Bacillus circulans
Bacillus firmus
Bacillus lentus
Bacillus stearothermophilus
Branhamella catarrhalis
Branhamella moraxella catarrhalis
Campylobacter fetus intestinalis
Campylobacter fetus jejuni
Chromobacterium violaceum
Citrobacter freundii
Clostridium perfringens
Corynebacterium diptheriae
Edwardsiella tarda
Enterobacter aerogenes
Escherichia coli
Flavobacterium rigense

Haemophilus influenzae
Kingella kingae
Klebsiella oxytoca
Klebsiella pneumoniae
Lactobacillus casei
Micrococcus luteus
Moraxella bovis
Moraxella urethralis
Mycobacterium avium
Mycobacterium chelonae
Mycobacterium fortuitum
Mycobacterium gordonae
Mycobacterium intracellulare
Mycobacterium kansasii
Mycobacterium tuberculosis
Mycoplasma fermentans
Mycoplasma genitalium
Mycoplasma hominis
Mycoplasma mycoides
Mycoplasma pulimonis
Neisseria flavescens
Proteus mirabilis

Proteus morganii
Proteus vulgaris
Providencia rettgeri
Pseudomonas aeruginosa
Pseudomonas cepacia
Pseudomonas fluorescens
Pseudomonas maltophila
Pseudomonas putida
Pseudomonas pyrrocinia
Pseudomonas solanacearium
Pseudomonas stutzeri
Salmonella enteritidis
Serratia marcescens
Shigella boydii
Shigella dysenteriae
Shigella sonnei
Spiroplasma apis
Staphylococcus aureus
Streptococcus pneumoniae
Streptococcus pyrogenes
Vibrio alginolyticus
Vibrio leiognathi
Alteromonas rubra

These species have been tested at 10⁶ organisms/ml of original water sample and are either not amplified, not detected, or both, with the exception of *Alteromonas rubra*. *Alteromonas rubra* will produce a weak "L" dot when 10⁶ organisms per ml are detected.

TABLE 3. Environmental Samples Sources and the Number That Inhibited PCR Amplification and Detection after Initial GuSCN Purification (PCR1) and after GuSCN Purification and BSA Treatment (PCR2).

	Number of Samples		
Sample Source	Total	Inhibited PCR1	Inhibited PCR2
Calorifier	10	3	0
Cooling tower	16	1	0
Shower	8	0	0
Faucet	26	0	0
Hot water system	9	0	0
Cold water system	3	0	0
Bulk hot water	2	0	0
Boiler house outlet	1	1	0
Water heater	1	1	0
Other	1	1	0
Total	87	7	0

treatment. The PCR detected the genus *Legionella* in all of the fifty-one culture-positive samples. Conversely, culturing detected the genus *Legionella* in fifty-one of the seventy-eight PCR-positive samples. The PCR detected *L. pneumophila* in thirty-six of the forty-three culture-positive samples. Culture detected the species *L. pneumophila* in thirty-six of the forty PCR-positive samples. Based on culture, the sensitivity of PCR for the genus *Legionella* and species *L. pneumophila* are 100% and 84%, respectively. Based on PCR, the sensitivity of culture for the genus *Legionella* and species *L. pneumophila* are 65% and 90%, respectively. The semiquantitative results (Table 4) indicate that PCR detected higher concentrations of organisms of the genus *Legionella* than culture while relatively comparable concentrations of *L. pneumophila* are detected by the two methods.

DISCUSSION

Samples that interfere with PCR amplification were detected by the absence of a positive IPC dot on reverse dot blots. While data indicating that PCR is inhibited by environmental samples may first appear alarming, reports [28–30] indicate that standard culture methods do not always detect *Legionella*. The extent of environmental samples that inhibit culture has not been determined, nor is it readily measurable. In some cases overgrowth of other bacteria may block the detection of *Legionella* and, while acid treatment may inhibit overgrowth of other bacteria, it may also reduce

TABLE 4. Comparison of PCR Semiquantitative Results to Culture.

a. Detection of the genus *Legionella*

		Culture		
	Negative	$\leq 10^3$	$= 10^3$	$\geq 10^3$
P Negative	9	0	0	0
C $\leq 10^3$	8	5	1	3
R $= 10^3$	9	2	3	2
$\geq 10^3$	10	7	4	24

b. Detection of the species *L. pneumophila*

		Culture		
	Negative	$\leq 10^3$	$= 10^3$	$\geq 10^3$
P Negative	40	4	0	0
C $\leq 10^3$	7	5	4	9
R $= 10^3$	0	0	2	6
$\geq 10^3$	0	0	0	10

sensitivity for *Legionella*. In other cases, *Legionella* may be viable but unculturable in environmental samples. In these aspects PCR presents an advantage over culture. While additional work is needed to eliminate the inhibitory effects of environmental samples on PCR amplification, tremendous progress has been made. Furthermore, since this protocol can generate results for a majority of environmental samples, the technique does have a practical, routine application. However, the incorporation of an internal positive control to monitor the efficiency of amplification is essential.

The observed differences in the sensitivities of PCR and culture found in the data generated by the Thames study may have several explanations. One explanation would be that culture may not be as sensitive for the detection of other species of *Legionella* as it is for *Legionella pneumophila*. When the samples containing *Legionella* species by PCR, but negative by culture, were PCR amplified using another PCR amplification system based on the 16s rRNA gene [68], twenty-two of the twenty-four (92%) samples remained PCR positive. These data would support the PCR results that the samples contain *Legionella*. Another explanation would be that PCR is detecting dead *Legionella* that cannot be detected by culture. While this PCR method does not discriminate between DNA from live and dead organisms, the membrane filtration does not select for DNA that is trapped within intact cells. Also, if this were the explanation, the sensitivity of PCR for the detection of the species *L. pneumophila* should be greater than culture, yet the sensitivities of the two methods for the species *L. pneumophila* are comparable. The existence of viable, nonculturable *Legionella* in the samples may also lead to such results. If this were the case, PCR may be an improvement over standard culture. The reasons for the discrepancies between culture and PCR are being investigated. It is essential that such investigations are performed to facilitate a better understanding of the PCR process.

The PCR is becoming accepted as a useful tool for detecting *Legionella* in the environment. The process of sample preparation, PCR amplification, and detection can be done in about five hours. Comparison studies on environmental samples between PCR, culture, and/or DFA have begun to demonstrate the value of the PCR method [13,48–52,69–73]. This method has proven useful for quickly identifying sources of outbreaks of disease where culture methods fail [13] and, with development of PCR amplification protocols that are capable of distinguishing among serotypes of *Legionella pneumophila* and other *Legionella* species, culture may no longer be necessary. The EnviroAmp™ Legionella kits have now been well tested against culture for a wide variety of sample types and function well. The test is easy to perform and requires little expertise in molecular biology techniques, and, therefore, is suitable for incorporation into the routine procedures of most microbiology laboratories. The vast improvement in speed for obtaining results is a major advantage over culture and may lead to a better understanding and possibly avoidance of outbreaks of legionellosis.

ACKNOWLEDGEMENT

The authors would like to thank C. Anthenien, R. LaFontaine, J. Raymond, C. Scanga and M. Zoccoli for their assistance in developing this procedure and C. Thio for her assistance.

REFERENCES

1 Mcdade, J. E., Shepard, C. C., Fraser, D. W., Tsai, T. F., Redus, M. A., Dowdle, W. R., and Laboratory Investigation Team. 1977. Legionnaires' disease: isolation of a bacterium and demonstration of its role in other respiratory disease. *N. Engl. J. Med.* 297: 1197–1203.

2 Brenner, D. J., Steigerwalt, A. G., and McDade, J. E. 1979. Classification of the Legionnaires' disease bacterium: *Legionella pneumophila,* genus novum, species nova, of the family Legionellaceae, familia nova. *Ann. Intern. Med.* 90: 656–658.

3 Bozeman, F. M., Humphries, and J. W., Campbell, J. M. 1968. A new group of rickettsia-like agents recovered from guinea pigs. *Acta Virol.* 12: 82–93.

4   Institutional outbreak of pneumonia. 1965. *MMWR*. 14: 265–286.

5   Thacker, S. B., Bennett, J. V., Tsai, T. F., Fraser, D. W., McDade, J. E., Shephard, C. C., Williams, K. H. Jr., Stuart, W. H., Dull, H. B., and Eickhoff, T. C. 1978. An outbreak in 1965 of severe respiratory illness caused by the Legionnaires' disease bacterium. *J. Infect. Dis.* 138: 512–519.

6   Herbert, G. A., Moss, C. W., McDougal, L. K., Bozeman, F. M., McKinney, R. M., Brenner, D. J. 1980. The rickettsia-like organisms TATLOCK (1943) and HEBA (1959): Bacterial phenotypically similar to but genetically distinct from *L. pneumophila* and the WIGA bacterium. *Ann. Internal. Med.* 92: 45–52.

7   Fang, G.-D., Yu, V. L., and Vickers, R. M. 1989. Disease due to the Legionellaceae (other than *Legionella pneumophila*). *Medicine.* 68: 116–132.

8   Dondero, T. J. Jr., Rendtorff, R. C., Mallison, G. F., Weeks, R. M., Levy, J. S., Wong, E. W., and Schaffner, W. 1980. An outbreak of Legionnaires' disease associated with a contaminated cooling tower. *N. Eng. J. Med.* 302: 365–370.

9   Hlady, W. G., Mullen, R. C., Mintz, C. S., Shelton, B. G., Hopkins, R. S., and Daikos, G. L. 1993. Outbreak of Legionnaire's disease linked to a decorative fountain by molecular epidemiology. *Am. J. Epidemiology.* 138: 555–562.

10 Stout, J. E., Yu, V. L., Muraca, P., Joly, J., Troup, N., and Tompkins, L. A. 1991. Potable water as a cause of sporadic cases of community-acquired Legionnaires' disease. *N. Eng. J. Med.* 326: 151–155.

11 Mahoney, F. J., Hoge, C. W., Farley, T. A., Barbaree, R. F., Benson, R. F., and Mc Farland, L. M. 1992. Community-wide outbreak of Legionnaires' disease associated with a grocery store mist machine. *J. Infect. Dis.* 165: 736–739.

12 Paszko-Kolva, C., Yamamoto, H., Shahamat, M., Sawyer, T. K., Morris, G., and Colwell, R. R. 1991. Isolation of Amoebae and *Pseudomonas* and *Legionella* spp. from eyewash stations. *Appl. Envir. Microbiol.* 57: 163–167.

13 Miller, L. A., Beebe, J. L., Butler, J. C., Martin, W., Benson, R., Hoffman, R. E., and Fields, B. S. 1993. Use of polymerase chain reaction in an epidemiologic investigation of Pontiac fever. *J. Infect. Dis.* 168: 769–772.

14 Fliermans, C. B., Cherry, L. H., Orrison, S. J., Smith, S. J., Tison, D. L., and Pope, D. H. 1981. Ecological distribution of *Legionella pneumophila. Appl. Envir. Microbiol.* 41: 9–16.

15 Fliermans, C. B., Cherry, W. B., Orrison, L. H., and Thacker, L. 1979. Isolation of *Legionella pneumophila* from non-epidemic-related habitats. *Appl. Envir. Microbiol.* 37: 1239–1242.

16 Stout, J. E., Yu, V. L., and Vickers, R. M. 1982. Ubiquitousness of *Legionella pneumophila* in the water supply of a hospital with endemic Legionnaires' disease. *New Engl. J. Med.* 306: 466–468.

17 Steele, T. W., Moore, C. V., and Sangster, N. 1990. Distribution of *Legionella longbeachae* serogroup 1 and other *Legionellae* in potting soils in Australia. *Appl. Environ. Microbiol.* 56: 2984–2988.

18 Guerrero, I., Genese, C., Hung, M. J., Paul, S., Ragazzoni, H., Brook, J., Finelli, L., Spitalny, K. C., Mojica, B. A., Mahoney, K. J., Heffernan, R. T., Kondracki, S. F., Morse, D. L., Cartter, M. L., Hadler, J., Rankin, J. T., and Groves, C. 1994. Update: outbreak of Legionnaires' disease associated with a cruise ship. *MMWR*. 43(31): 574–575.

19 Winn, W. C., Jr. 1988. Legionnaires' disease: historical perspective. *Clin. Microbiol. Reviews.* 1: 60–81.

20 World Health Organization. 1990. Epidemiology, prevention and control of legionellosis: memorandum from a WHO meeting. *Bull. World Health Org.* 68: 155–164.

21 Reingold, A. L., Thomason, B. M., Brake, B. J., Thacker, L., Wilkinson, H. W., and Kuritsky, J. N. 1984. *Legionella* pneumonia in the United States: the distribution of serogroups and species causing human illness. *J. Infect. Dis.* 149: 819.

22 Rowbotham, T. J. 1980. Preliminary report on the pathogenicy of *Legionella pneumophila* for fresh-water and soil amoebae. *J. Clin. Pathol.* 33: 1179–1183.

23 Tyndall, R. L. and Dominique, E. L. 1982. Cocultivation of *Legionella pneumophila* and free-living amoebae. *Appl. Environ. Microbiol.* 44: 954–959.

24 Skinner, A. R., Anand, C. M., Malic, A. and Kurtz, J. B. 1983. *Acanthamoebae* and environmental spread of *Legionella pneumophila. Lancet.* i: 289–290.

25 Kilvington, S. and Price, J. 1990. Survival of *Legionella pneumophila* within cysts of *Acanthamoeba polyphaga* following chlorine exposure. *J. Appl. Bacteriol.* 68: 519–525.

26 Muraca, P. W., Yu, V. L., and Stout, J. E. 1988. Environmental aspects of Legionnaires disease. *JAWWA.* 80: 78–86.

27 Edelstein, P. H. 1981. Improved semi-selective medium for isolation of *Legionella pneumophila* from contaminated clinical and environmental specimens. *J. Clin. Microbiol.* 14: 298–303.

28 Shahamat, M., Paszko-Kolva, C., Keiser, J., Colwell, R. R. 1991. Sequential culturing method improves recovery of *Legionella* spp. from contaminated environmental samples. *Intern. J. Med. Microbiol.* 275: 312–319.

29 Sanden, G. N., Morrill, W. E., Fields, B. S., Breiman, R. F. and Barbaree, J. M. 1992. Incubation of water samples containing amoebae improves detection of Legionellae by the culture method. *Appl. Environ. Microbiol.* 58: 2001–2004.

30 Hussong, D., Colwell, R. R., O'Brien, M., Weiss, E., Pearson, A. D., Weiner, R. M., and Burge, W. D. 1987. Viable *Legionella pneumophila* not detectable by culture on agar media. *Bio/Technology.* 5: 947–950.

31 Grimont, P. A. D., Grimont, F., Desplaces, N., and Tchen, P. 1985. DNA probe specific for *Legionella pneumophila. J. Clin. Microbiol.* 21: 431–437.

32 Edelstein, P. H. and Edelstein, M. A. C. 1989. Evaluation of the Merifluor-*Legionella* immunofluorescent reagent for identifying and detecting 21 *Legionella* species. *J. Clin. Microbiol.* 27: 2455–2458.

33 Pasculle, A. W., Veto, G. E., Krystofiak, S., McKelvey, K., and Vrsalovic, K. 1989. Laboratory and clinical evaluation of a commercial DNA probe for the detection of *Legionella* spp. *J. Clin. Microbiol.* 27: 2350–2358.

34 Vickers, R. M., Stout, J. E., and Yu, V. L. 1990. Failure of a diagnostic monoclonal immunofluorescent reagent to detect *Legionella pneumophila* in environmental samples. *Appl. Envir. Microbiol.* 56: 2912–2914.

35 Ezaki, T., Hashimoto, Y., Yamamoto, H., Lucida, M. L., Liu, S. L., Kusunoke, S., Asano, K., and Yabuuchi, E. 1990. Evaluation of the microplate hybridization method for rapid identification of *Legionella* species. *Eur. J. Clin. Micro. Infect. Dis.* 9: 213–217.

36 Grimont, F., Lefevre, M., Ageron, E., and Grimont, P. A. D. 1989. rRNA gene restriction patterns of *Legionella* species: a molecular identification system. *Res. Microbiol.* 140: 615–626.

37 Tompkins, L. S., Troup, N. J., Woods, T., Bibb, W., and McKinney, R. M. 1987.

Molecular epidemiology of *Legionella* species by restriction endonuclease and alloenzyme analysis. *J. Clin. Microbiol.* 25: 1875–1880.

38 Saiki, R. K., Gelfand, D. H., Stoffel, S., Scharf, S. J., Higuchi, R., Horn, G. T., Mullis, K. B., and Erlich, H. A. 1988. Primer-directed enzymatic amplification of DNA with a thermostable DNA polymerase. *Science.* 239: 487–491.

39 Saiki, R. K., Scharf, S., Faloona, F., Mullis, K. B., Horn, G. T., Erlich, H. A., and Arnheim, N. 1985. Enzymatic amplification of β-globin genomic sequences and restriction site analysis for diagnosis of sickle cell anemia. *Science.* 230: 1350–1354.

40 Scharf, S. J., Horn, G. T. and Erlich, H. A. 1986. Direct cloning and sequence analysis of enzymatically amplified genomic sequences. *Science.* 223: 1076–1078.

41 Mullis, K. B., and Faloona, F. A. 1987. Specific synthesis of DNA in vitro via a polymerase-catalyzed chain reaction. In: Wu, R. ed. *Methods in Enzymology, Vol. 155.* San Diego, CA, Academic Press, pp. 335–350.

42 Starnbach, M. N., Falkow, S., and Tompkins, L. S. 1989. Species-specific detection of *Legionella pneumophila* in water by DNA amplification and hybridization. *J. Clin. Microbiol.* 27: 1257–1261.

43 Mahbubani, M. H., Bej, A. K., Miller, R., Haff, L., DiCesare, J., and Atlas, R. M. 1990. Detection of *Legionella* with polymerase chain reaction and gene probe methods. *Molecular and Cellular Probes.* 4: 175–187.

44 Bej, A. K., Mahbubani, M. H., Miller, R., DiCesare, J., Haff, L. and Atlas, R. M. 1990. Multiplex PCR amplification and immobilized capture probes for detection of bacterial pathogens and indicators in water. *Molecular and Cellular Probes.* 4: 353–365.

45 Bej, A. K., Mahbubani, M. H., and Atlas, R. M. 1991. Detection of viable *Legionella pneumophila* in water by Polymerase Chain Reaction and Gene Probe Methods. *Appl. Envir. Microbiol.* 57: 597–600.

46 Dreyfus, L. A. 1989. Molecular cloning and expression in *Escherichia coli* of the *rec*A gene of *Legionella pneumophila*. *J. General Microbiol.* 135: 3097–3107.

47 Somerville, C. C., Knight, I. T., Straube, W. L., and Colwell, R. R. 1989. Simple, rapid method for direct isolation of nucleic acids from aquatic environments. *Appl. Envir. Microbiol.* 55: 548–554.

48 Bowman, E. K. and Tyndall, R. 1993. Evaluation of cooling tower water samples with the EnviroAmp™ polymerase chain reaction (PCR) *Legionella* test kit. Abstract for the 93rd ASM General Meeting.

49 Hackman, B. and Plouffe, J. 1993. Polymerase chain reaction for *Legionella* in environmental water samples. Abstract for the 93rd ASM General Meeting.

50 Martin, W. T., Fields, B. S., and Huntwagoner, L. C. 1993. Comparison of culture and polymerase chain reaction to detect *Legionellae* in environmental samples. In: Barbaree, J. M., Breiman, R. F., and Dufour, A. P., eds. Legionella *Current Status and Emerging Perspectives.* ASM, Washington, D.C.

51 Matthias, M., Kissel, K., Scrimuang, S., Doeberitz, M. K., and Sonntag, H.-G. 1994. Comparisons of polymerase chain reaction and conventional culture for the detection of *Legionella* in hospital water samples. *J. Appl. Bacteriol.* 76: 216–225.

52 McCarty, S., Burden, J., Wheeler, T., Burton, B., and Smith, J. 1994. Evaluation of the EnviroAmp Legionella Kits for the detection of *Legionellaceae* in environmental samples. *Amplifications.* The Perkin-Elmer Corporations.

53 Castillo, I., Bartolome, J., Quiroga, A., and Carreno, V. 1992. Comparison of several PCR procedures for detection of serum HCV-RNA using different regions of the HCV genome. *J. Virol. Meth.* 36: 71–80.

54 T. Picone et al., unpublished data.

55 MacDonell, M. T. and Colwell, R. R. 1987. The nucleotide sequence of the 5S rRNA from *Legionella pneumophila. Nuc. Acids Res.* 15: 1335.

56 Engleberg, N. C., Carter, C., Weber, D. R., Cianciotto, N. P., and Eisenstein, B. I. 1989. DNA sequence of *mip*, a *Legionella pneumophila* gene associated with macrophage infectivity. *Infec. Immun.* 57: 1263–1270.

57 Cianciotto, N. P., Bangsborg, J. M., Eisenstein, B. I., and Engleberg, N. C. 1990. Identification of *mip*-like genes in the genus *Legionella. Infec. Immun.* 58: 2912–2918.

58 Cianciotto, N. P., Eisenstein, B. I., Mody, C. H., and Engleberg, N. C. 1990. A mutation in the *mip* gene results in an attenuation of *Legionella pneumophila* virulence. *J. Infectious Diseases.* 162: 121–126.

59 Wolters, J. and Erdmann, V. A. 1988. Compilation of 5S rRNA and 5S rRNA gene sequences. *Nuc. Acids. Res.* 16 (Supplement): r 1–r 70.

60 Chumakov, K. M., Tartakovsky, I. S., Ogarkova, O. A., and Prozorovsky, S. V. 1986. Use of the 5S ribosomal RNA nucleotide sequence analysis for the study of the phylogeny of the genus *Legionella. Molekuliarnaia Genetika, Mikrobiologia, I Virosolog.* 7: 38–40.

61 Picone, T. P. unpublished data.

62 Blake, E., Mihalovich, J., Higuchi, R., Walsh, P. S. and Erlich, H. 1992. Polymerase chain reaction (PCR) amplification and human leukocyte antigen (HLA)-DQ alpha oligonucleotide typing on biological evidence samples: casework experience. *J. Forensic Sci., JFSCA.* 37: 700–726.

63 Higuchi, R. and Kwok, S. 1989. Avoiding false positives with PCR. *Nature.* 339: 237–238.

64 Kwok, S. 1990. Chapter 17. Procedures to minimize PCR-product carry-over, and Orrega, C. 1990. Chapter 54. Organizing a laboratory for PCR work. In: Innis, M. A., Gelfand, D. H., Sninsky, J. J. and White, T. J., eds. *PCR Protocols: A Guide to Methods and Applications.* Academic Press, Inc., San Diego, CA.

65 Longon, N., Berninger, N. S., and Hartley, J. L. 1990. Use of Uracil DNA glyco-sylase to control carry-over contamination in polymerase chain reactions. *Gene.* 93: 125–128.

66 Saiki, R. K., Walsh, P. S., Levenson, C. H., and Erlich, H. A. 1989. Genetic analysis of amplified DNA with immobilized sequence-specific oligonucleotide probes. *Proc. Natl. Acad. Sci. USA.* 86: 6230–6234.

67 Morris, G. K. and Shelton, B. K. 1991. *Legionella* in environmental samples: hazard analysis and suggested remedial actions. Pathcon Technical Bulletin 1.3, Pathogen Control Associates, Inc. Norcross, GA.

68 Dodge, D. E., Dare, A., and Young, K. 1993. Coamplification and specific detection of three causative agents of acute pneumonia using the polymerase chain reaction. Poster for the 93rd ASM General Meeting.

69 Danielson, R. E., Lindquist, D., Wong, J. D. *Legionellae* detection in a building associated with a Legionnaires' disease outbreak. Submitted for publication.

70 Palmer, C. J., Tsai, Y.-L., Paszko-Kolva, C., Mayer, C., Sangermano, L. R. 1993. Detection of *Legionella* species in sewage and ocean water by polymerase chain reaction, direct fluorescent antibody, and plate culture methods. *Appl. Environ. Microbiol.* 59: 3618–3624.

71 Palmer, C. J., Tsai, Y.-L., Paszko-Kolva, C., Sangermano, L. R., Bonilla, G. F., Roll, B., Fujioka, R. S. 1993. Detection of *Legionella* species in sewage and ocean water in California and Hawaii. Water Environment Federation 66th Annual Conference & Exposition.

72 Palmer, C. J., Bonilla, G. F., Roll, B., Paszko-Kolva, C., Sangermano, L. R., Fujioka, R. S. Detection of *Legionella* species in reclaimed water and air using the EnviroAmp PCR detection kit and direct fluorescent antibody staining. Submitted for publication.

73 Fricker, E. J. and Fricker, C. R. (In press). Detection of legionellas in water samples using the polymerase chain reaction. *Journal of Applied Bacteriology.*

CHAPTER 9

DNA Amplification Techniques in Fossilized Samples

RAÚL J. CANO[1]

INTRODUCTION

THE isolation and characterization of fossil DNA, until recently [27], was considered unattainable because the methodologies for extracting minute quantities of partially degraded DNA and their subsequent enzymatic amplification were not available. With the advent of the polymerase chain reaction [38], a new analytical tool became available for the molecular study of fossils. It is now possible to conduct molecular studies of extinct organisms utilizing their DNA to unravel biological and evolutionary questions.

There is already a body of scientific evidence that supports the use of DNA from extinct animals and plants for phylogenetic studies. Poinar and Hess [48] reported ultrastructural evidence indicating that 40-million-year-old (myo) Baltic amber contained partially preserved insect tissue, including nuclei, mitochondria, muscle fibers, and endoplasmic reticulum. Higuchi and coworkers [27,28] demonstrated that remains of a mammoth and the extinct species, the quagga, contained fragments of the original DNA. Pääbo [43,44] reported the extraction of clonable DNA from a 2,400-year-old mummy of a child. Subsequent DNA analysis revealed fragments measuring approximately 3.4 kilobase pairs (Kbp). Thomas et al. [58] isolated DNA from hair found in a century-old, untanned hide and a piece of dried muscle collected from an extinct marsupial wolf. This DNA was later enzymatically amplified by PCR and phylogenetic studies were made.

[1]Biological Sciences Department, California Polytechnic State University, San Luis Obispo, CA 93407, U.S.A.

More recently Golenberg et al. [19] isolated and analyzed *Magnolia* chloroplast DNA from a Miocene *Clarkia* deposit dated 17–20 myo. Cano et al. [4,6] isolated and characterized DNA from the extinct bee *Proplebeia dominicana* in 25–40 myo Dominican amber. DeSalle et al. [14] employed DNA extracted from fossil termites to resolve phylogenetic relationships between the termites, cockroaches, and mantids. Cano et al. [7] extracted DNA from a 120–135 myo nemonychid weevil in Lebanese amber and showed by nucleotide sequence alignments and phylogenetic inference analyses that the fossil weevil was most closely related to the extant nemonychid weevil *Lecontellus pinicola*. Poinar et al. [47] used DNA sequences from the extinct legume *Hymenaea protera* in Dominican amber for a biogeographical study in which they showed that the extinct *H. protera* was most closely related to the extant African species *H. verrucosa*, as morphological studies suggested. Finally, Cano et al. [5] used DNA sequences from 25–40 myo *Bacillus* spp. in Dominican amber inclusions to study a symbiotic relationship between *Bacillus* and the now extinct stingless bee *Proplebeia dominicana*.

The value of fossil evidence is that it may demonstrate the condition of taxa before evolutionary divergence obscured phylogenetic relationships [15]. Because they are older, ancient fossil DNA sequences should be less divergent than extant sequences and should, therefore, have value for relating more derived extant taxa. When compared with extant DNA, ancient DNA sequences may also provide an insight into the pattern of molecular evolutionary change through time. Fossil DNA has been used to answer evolutionary questions among organisms [14], detect the presence of pathogens in museum specimens [46], study the origin of Pacific Island populations [23], and study spatial and temporal distribution of populations [59].

An interesting question that can be addressed with fossil DNA is the "Molecular-Clock Hypothesis." Fossil-DNA sequence data can be used for estimating the rate and pattern of molecular change through time [37,42]. To study this pattern, it might be possible to compare typical pairwise distances derived from nucleotide sequence data measured between extant genera with the distances measured between the fossils and unrelated extant taxa.

Much of what is known about extinct organisms comes from traits that are not preserved in the fossil record. Mating strategies, sexual dimorphism, gestational rates, feeding behaviors, coloration, and other soft-tissue characteristics may be derived traits that directly influence placement in phylogenetic analysis. Sample size and incompleteness of specimens also present problems in phylogenetic analysis of fossilized organisms. Because of the biases in the fossil record, there are generally few samples on which to base such analyses, and quite often placement of taxa is based on the traits of single bones, museum skins, or teeth.

Until recently, morphological analysis based on skeletal remains was the only tool available for scientists to determine proximity of sister groups as well as ancestor-descendant relationships for extinct fossil organisms. The proposed lineages for dinosaurs as a group are based on careful analysis of the traits that are revealed in the bones and other "hard" parts preserved in the fossil record. Much information is contained in the characteristics of the bones, and by comparing micro- and macrostructures with those of living animals, many traits of extinct organisms have been inferred [8,11,12,51–53,61]. Much information, however, is also lost in the decay of soft tissues and cells, which would carry biochemical information, internal organization, and perhaps indications of thermoregulatory capacities.

If it can be shown, however, that DNA or other biomolecules can be preserved, at least in small pieces in the remains of extinct organisms, then this information may not be lost after all. By targeting specific gene sequences, it may be possible to deduce biochemical characteristics. Also, through comparative studies of sequences from various extant taxa, it is possible to estimate the extent of evolutionary divergence [30,41,60]. By comparing the amount and type of these variations, one could estimate how quickly some DNA "evolves" relative to other segments, or which bases or genes have the most flexibility or are more conserved over time. The ability to isolate DNA directly from extinct, and presumably ancestral lineages, would allow the direct verification of currently extrapolated conclusions and give concrete validation to such estimates as rate of changes and direction or polarity in base-pair changes [13,16,42].

The compilation of these data would yield much understanding of the physiology of extinct creatures, including dinosaurs, that is not attainable through other methods. Also, it would provide a much clearer picture of genetic change over time and the mechanics behind "evolution."

STRATEGIES OF ANALYSIS

The explosion in the field of biotechnology has made areas of study available to paleontology that were never before possible. In terms of the analysis of DNA, the single most important technology is the polymerase chain reaction. This exponential amplification produces enough copies of the target strand of DNA to be manipulated and analyzed through standard molecular techniques, such as cloning and enzymatically directed sequencing [55]. Coupled with new and refined techniques for extraction of biomolecules tightly adhered to matrices, this technology has become a powerful tool for analysis in molecular paleontology.

Analytical software is available (e.g., CLUSTAL [25,26], FASTA [36], and GDE [Steve Smith, personal communication]) that allows the sequences obtained by the above methods to be matched against homologous sequences from other species that have been entered into a data bank. Statistical analyses can then be performed and estimates to relatedness and genetic distance can be obtained [32]. Phylogenetic trees based on sequence data can be constructed using software packages, such as MEGA [33], PAUP [57], and PHYLIP [18]. This allows for the objective placement of an organism within the framework of known taxa (Figure 1). It also allows any modern DNA, which may be contaminating ancient tissues, to be characterized and possibly recognized.

SELECTIONS OF GENE SEQUENCES FOR ANALYSIS OF FOSSIL DNA

When working with DNA putatively obtained from fossils, the selection of gene sequences for amplification and/or analysis is a crucial step. In the case of extinct organisms for which there is no direct living representative, the genomes of the closest living relatives (based on morphological analysis) are examined for conserved sequences. When selected regions of genes for these taxa are compared, homologous sequences can be identified. Regions of homology, where at least fifteen bases are identical between the two groups, are good places to start when designing primer molecules.

The size of the amplified target sequence (amplicon) is also of importance. Generally, when designing primers to amplify DNA segments from fossils, it is best to think small. The chances for successful amplifications of fossil sequences increase as the size of the amplicon decreases. As a general rule, it is recommended that the selected primer pair amplify a region of the desired gene to measure ≤ 200 bp (Figure 2). As the fossil DNA becomes damaged and degraded, the resulting fragment length becomes smaller. Thus, amplification of small DNA segments will be more successful than that of larger segments [24]. Table 1 illustrates the results of a study conducted in the laboratory. The study aimed at demonstrating the reproducibility of DNA extraction from amber inclusions (dated 25–40 million years old) of the extinct bee *Proplebeia dominicana*, the extant bee *Plebeia frontalis,* and their corresponding *Bacillus* symbionts. The amplification products are illustrated in Figure 2.

The chances for successful amplification are increased if the target gene sequence is present in multiple copies within each cell. Nuclear DNA sequences of ribosomal constituents, such as 18s and 28s rDNA, are often used in such studies. Mitochondrial DNA sequences are also good candidates because, not only are there several to thousands of mitochondria per cell, the complete mitochondrial genome for many taxa have been sequenced and entered into data banks and are available for comparative studies. Some primer pairs are described in Table 2.

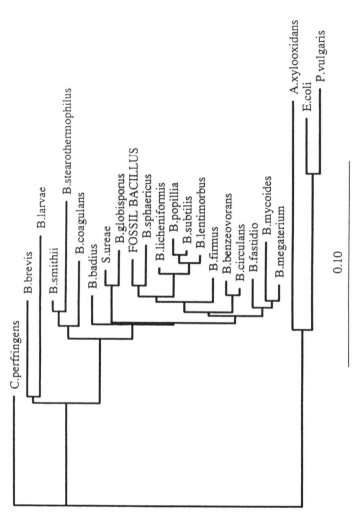

Figure 1 Phylogenetic tree, constructed using the Maximum Likelihood algorithm, that includes extant, and putative fossil bacterial sequences. The DNA was extracted from both extant and extinct bees and amplified using *Bacillus* specific primers. This tree supports the hypothesis that the putative ancient sequence is most closely related to extant *Bacillus*.

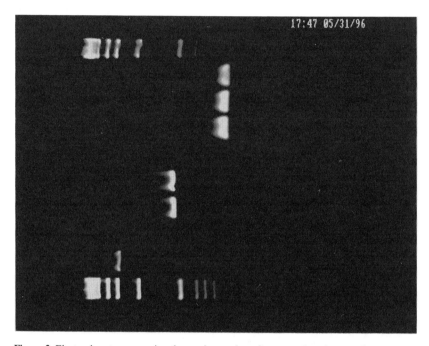

Figure 2 Electrophoretogram, using three primer pairs, of extant and ancient amplicons. Lanes 1 and 10: 1 Kb ladder (Gibco-BRL, Gaithersburg, MD). Lane 2: results of an amplification using DNA from the abdominal tissue from an extant bee using a primer pair that amplifies a 1400 bp segment of the bacterial 16s rRNA gene. Lane 3: results of an amplification using DNA from a 25–40-million-year-old amber-entombed bee using the same primer pair as in Lane 2. Lane 4: results of an amplification using DNA from the abdominal tissue from an extant bee using a primer pair that amplifies a 550 bp segment of the bacterial 16s rRNA gene. Lanes 5 and 6: results of an amplification using DNA from a 25–40-million-year-old amber-entombed bee using the same primer pair as in Lane 4. Lane 7: results of an amplification using DNA from the abdominal tissue from an extant bee using a primer pair that amplifies a 206 bp segment of the bacterial 16s rRNA gene. Lanes 8 and 9: results of an amplification using DNA from a 25–40 million-year-old amber-entombed bee using the same primer pair as in Lane 7.

188

TABLE 1. Amplification Efficiency of Fossil and Extant DNA Samples.

Sample ID	BCA[a]	Int3[b]	16s[c]	NS2/19[d]	NS1/4[e]
P. dominicana	6/16[f]	7/16	0/16	6/16	0/16
Plebeia frontalis	6/6	6/6	4/6	6/6	4/6
Bacillus sphaericus	4/4	4/4	4/4	0/4	0/4
Bacillus subtilis	4/4	4/4	4/4	0/4	0/4

[a] The primer pair BCA341F/BCA871R amplifies a 530 bp segment of Bacillus spp. 16s rRNA (see Reference [5]). [(BCA341F: 5′-TACgggAggCAgCAgTAgggAAT-3′), (BCA871R: 5′-TACTCCCCAggCggAgTgCTTAAT-3′)].

[b] BCAInt3/BCA871R amplify a 336 bp segment of Bacillus spp. 16s rRNA. The sequence of BCAInt3 is 5′-TgCCAgCAgCCCgCggTAT-3′.

[c] This primer pair amplifies ~1400 bp segment of eubacterial 16s rRNA. (16sH: 5′-TNANACATgCAAgTCgAICg-3) corresponds to positions 49–68 of E. coli 16s rRNA and the reverse primer (16sL: 5′-ggYTACCTTgTTACgACTT-3′).

[d] The primer pair NS2/NS19 amplifies a 177–200 bp fragment of 18s rRNA. [(NS2:5′-ggCTgCTgg-CACCAgACTTgC-3′), (N19: 5′-CCggAgAAggAgCCTgAgAAAC-3′)].

[e] The primer pair NS1/NS4 amplifies ~1200 bp fragment of 18s rRNA [(NS1:5′-gTAgTCATATgCT-TgTCTC-3′) (NS4: 5′-CTTCCgTCAATTCCTTTAAg-3′)].

[f] The numerator represents the number of successful amplifications using the primer pair as determined by an amplicon of the expected size and all controls yielding the appropriate results. The denominator represents the total number of samples tested.

TABLE 2. Primer Pairs for the Amplification of Eukaryotic rRNA Genes.

Code	Sequence	Gene	Size
NS1[a]	gTAgTCATATgCTTgTCTC	18s rRNA	555 bp
NS2	ggCTgCTggCACCAgACTTgC	18s rRNA	
NS3[a]	gCAAgTCTggTgCCAgCAgCC	18s rRNA	597 bp
NS4	CTTCCgTCAATTCCTTTAAg	18s rRNA	
16Sbr[b]	CCggTCTAACTCAgATCACgT	16s rRNA	500–650 bp
16Sar	CgCCTgTTTAACAAAAACAT	16s rRNA	
12Sbi[b]	AAgAgCgACgggCgATgTgT	12s rRNA	411–318 bp
12Sai	AAACTAggATTAgATACCCTATTAT	12s rRNA	

[a] White et al. [64].
[b] Simon et al. [56].

For phylogenetic significance, it is desirable that the selected homologous primer sequences for known taxa flank regions of relatively high variability. This allows for better definition of phylogenetic placement than if there are relatively few changes across a broad range of taxa. Also, it is easier to tell if there is contamination with modern DNA. If DNA from ancient samples can be obtained and amplified, then analysis can reveal if any base-pair changes in the sequences from ancient materials are intermediate between the modern taxa being used for comparison.

PRESERVATION POTENTIAL OF BIOMOLECULES

It is a commonly held belief, based on experimental evidence as well as extrapolated predictions based on studies of DNA in aqueous solution, that nucleic acids do not survive in fossil remains on a geological time scale [35]. These assumptions, however, are being challenged by researchers who are continually pushing back the age for identification and recovery of DNA and proteins obtained from fossils preserved under rare and specific conditions [4–7,10,14,19,21–23,27,28,39,40,43–45].

The double stranded, helical structure of DNA is more resistant to damage than single-stranded RNA [35], but its structure and chemistry make it susceptible to certain types of damage over time. Conversion of bases through hydrolytic deamination (guanine changes to xanthine, cytosine to uracil or its derivatives) and depurination (removal of the bases guanine and adenine from the sugar-phosphate backbone) affect the informational content of the molecule. Exposure to oxygen-free radicals or UV radiation also damages DNA strands [17]. Mechanisms have evolved in living organisms for repairing such DNA damage as it occurs, maintaining genetic information, and preventing accumulation or errors [20]. With the death of the organism, this self-repair process stops, while enzymatic attack and exposure to water, oxygen, and ultraviolet radiation continue with advancing decay. There are rare cases, however, where DNA is protected from such damage.

Exposure to water is probably the single most destructive force acting on the DNA molecule. Water has been shown to initiate strand breaks by attacking the base-sugar bonds. Where the base is lost, the chain is weakened and eventually cleaved [17,35]. Given these facts, a crucial step in the preservation of DNA is relatively rapid dehydration of tissues. One way this occurs is through entrapment of organisms in amber-forming resins.

Amber is an amorphous, polymeric glass, with mechanical, dielectric, and thermal features common to synthetic polymers [63]. It originates from the resin of woody plants, and is commonly recognized as sticky odoriferous "pitch." Natural resins are complex mixtures of terpenoid compounds, acids, alcohols, and saccharides secreted from parenchymal cells, some of which have preservative and antimicrobial properties [34,49,50]. Resins are not restricted to the conifers but occur in a wide range of flowering plants [34]. Through the aging processes of oxidation and polymerization, the resin becomes harder and ultimately forms the gemstone known as amber. The preservative properties of amber make it a suitable source of tissue with extractable DNA, from which genetic studies can be conducted [4,5,7,14,47].

What makes amber such a good preservative of DNA? Studies conducted on the trunk resin of the tree *Agathis australis* may provide part of the answer. First, the sugars arabinose, galactose, and sucrose are present in such resins. High concentrations of these sugars in the resin would make the resin hyperosmotic to the cell by drawing water out and achieving tissue dehydration. Under water-free conditions, biochemical reactions, including those involved in the degradation of nucleic acids and proteins, are inhibited. Microbial activity that results in the degradation of cellular components is also halted because there is not sufficient water to carry out microbial metabolism.

Second, alcohols such as fenehyl and communol and terpenes such as alpha-pinine, limonene, and dipentene may act as fixatives to preserve tissue. Evidence of such preservative properties can be seen in the electron photomicrographs of Poinar and Hess [50] in which they show evidence of chromatin, endoplasmic reticulum, and mitochondria of a 40 myo midge fly in Baltic amber. Additionally, one of the oxygenated derivatives of terpene hydrocarbons is aldehyde, which may also serve as a fixative of embedded tissue.

Effective dehydration can also occur with the removal of DNA from solution. This process occurs through adsorption of DNA onto mineral surfaces [54]. Hydroxyapatite is known to have a very strong binding affinity for DNA [1,30] and this component is, of course, the mineral that predominates in bone. Removal from solution through adsorption protects the molecule from attack by hydronium ions.

Another consideration in the long-term preservation of DNA is the pH of the environment. Acidic environments may increase the rate of degradation of this molecule as H^+ ions can attack the OH groups of the sugars and the nitrogenous bases, contributing to strand breakage. Bone also sets up an alkaline environment (hydroxyapatite is a basic compound), which can favor the preservation of DNA. However, Lindahl [35] claims that in the vicinity of 7.4, variations in pH do not seem to be a major factor in the degradation of DNA.

Oxidation is another source of DNA damage, and removing DNA from water, as in amber or bone, protects the molecule from oxidative attack.

Oxygen, in its molecular state, does not attack DNA, but rather it is the formation of oxygen-free radicals that attack the nitrogenous bases. Oxidative attack would be rapid at first, but then would level off [44]. It is proposed that chelation of copper or other metal ions [17,54] enhances preservation by contributing to a reducing environment and compensating for the production of oxygen-free radicals.

Exposure to ultraviolet light also causes extensive damage and degradation of DNA, and rapid burial of an organism is important to minimize the consequences of UV damage to DNA. Rapid burial is implied in the preservation of fossils, such as fossil bones. It is assumed that predation, bloat, bacterial decay, scavenging, and other taphonomic processes seen today were equally active in the breakdown of organic remains during prehistoric times. To avoid total disintegration of remains by these forces, burial must have occurred relatively soon after death. This is particularly true when skeletons are found fully articulated. The assumption is made that burial occurred before the soft tissues like ligaments, muscles, and skin, which hold the bones together, had undergone complete decay.

PROBLEMS OF WORKING WITH "FOSSIL" BIOMOLECULES

The extreme sensitivity of PCR, which opens the door to the direct analysis of DNA obtained from ancient materials, also poses the most complications. The fact that PCR technology can amplify as little as one molecule of DNA means that minute amounts of contaminating DNA from modern sources, such as bacteria, soil fungi, or human skin cells can also be amplified. Indeed, any such modern contaminant would probably be amplified preferentially over ancient target molecules owing to the probable state of degradation of the latter. It is for this reason the selection of primer molecules used in amplification is such a crucial step because careful design can decrease or eliminate spurious amplification of contaminating DNA. Through studies of published sequences of extant species, it is desirable to build primer molecules from regions that would prevent the amplification of DNA from the most common sources of contamination. Also, it becomes very important to run several environmental controls at each step of the isolation and amplification process. If the gene sequences chosen for amplification flank regions of variability, or regions containing insertion or deletions, then analysis of sequence data obtained from PCR amplification of ancient targets makes contamination by modern DNA much easier to detect.

Limiting access to ancient DNA laboratories and equipment reduces potential sources for contamination. Frequent washing of lab surfaces with a 10% bleach solution, and continuous exposure of surfaces and reagents to UV light when not in use also reduces the potential for contamination, because UV light is known to cross-link DNA strands, thus making them unavailable for amplification by PCR. Keeping labs used in ancient DNA work separated from any labs used in modern analyses is another important requirement. In addition, separating areas for extraction of DNA from areas designated for setting up PCR reactions also minimizes the possibility of contamination.

Ultimately, however, the proof of the authenticity of any DNA presumably obtained from ancient materials comes from careful analysis of sequence data. If phylogenetic analysis of the sequences does not agree with predicted relationships based on morphological data, particularly with species such as dinosaurs that leave no modern representatives, then the DNA data must be carefully reevaluated. Also, the analysis of at least two different genes or gene regions should be done, and the results of both should show similar or identical phylogenies, before any claims can be made regarding the sources of the DNA.

EXTRACTION AND ENZYMATIC AMPLIFICATION OF DNA FROM FOSSILS

The success of any enzymatic amplification by PCR is largely dependent on the preparation of the sample (i.e., DNA extraction) and the care taken to avoid contamination with extraneous DNA. Extraction of DNA from fossils presents challenges that are only approximated by the extraction of DNA from clinical and forensic specimens. As the DNA in these samples are both partially degraded and potentially contaminated with DNA from extant organisms, it is imperative that suitable methods be developed for processing fossil specimens and extracting DNA. Generally, methods, such as the well known phenol-chloroform method, for extracting genomic DNA are not well suited for fossil materials because they are relatively inefficient with low concentrations of DNA. Hagelberg and Clegg [23], however, have successfully isolated and amplified mitochondrial and genomic DNA using a modification of the standard phenol-chloroform assay.

The Chelex 100 extraction method has also been shown to yield amplifiable quantities of DNA from both forensic [62], museum skins [59], and amber-entombed specimens [4,7]. The DNA extracted by this method, however, may degrade after several months of storage at $-20°C$ and can no longer be amplified by PCR [3].

The use of chaotropic salts like NaI or guanidinium thiocyanate, which promote the binding of DNA to silica particles, are very useful for extracting DNA from fossils, including amber inclusions, ancient bones, and museum specimens [3,29]. Once the DNA is bound to the silica particles, it can be washed with 70% ethanol (or 70% ethanol in 0.1 M NaCl) to remove any PCR inhibitors that might be present in the sample [44] and eluted with small volumes (10–100 μl) of low ionic strength buffers (e.g., TE buffer or distilled water), thus permitting the concentration of small amounts of DNA.

LABORATORY DESIGN

Enzymatic amplification of DNA by PCR can be performed with as little as a single DNA molecule. Therefore, contaminating DNA, if it shares homologies with primer sequences and the target DNA could be amplified. It is imperative that proper laboratory protocols be implemented in order to validate all DNA isolations and amplifications. For this reason the following validation procedures should be implemented: (1) thorough training of the personnel involved in the collection and processing of specimens; (2) separate extinct- and extant-samples facilities; (3) separate pre- and post-PCR work areas to reduce product carryover contamination; and (4) stringent, aseptic techniques and work-space DNA and biological decontamination. This latter procedure should include weak positive controls to reduce the risk of carryover contamination, routine monitoring of environmental contamination with DNA by performing DNA-free PCRs using a panel of commonly used primers, multiple replicates of sample preparation and DNA amplifications, and statistical evaluation of the data by sequence alignment and phylogenetic inference analyses.

The basic PCR laboratory layout should consist of three areas: (1) reagent preparation area (Area 1); (2) sample preparation area (Area 2); and (3) PCR analysis area (Area 3). Additionally, separate laboratories should be used whenever possible for the processing of extinct and extant taxa.

DNA extractions should be carried out in areas designated solely for the extraction of extinct DNA. Separate areas (in two different rooms if possible) should be used to process extinct and extant specimens. Dedicated thermal cyclers and pipetting devices (with aerosol-resistant tips) should also be used to reduce the risk of cross contamination. Similarly, extant DNA material must be processed in a site remote from that employed for extraction of extinct DNA. The preparation of reagent mixes for the polymerase chain reaction should also take place within dead air boxes in designated areas away from sample preparation and PCR evaluation areas. Sample carryover can also be avoided by the use of filter-sterilized and/or autoclaved reagents and by decontamination of utensils and work areas with 10% bleach followed by UV irradiation. Electrophoresis must be carried out in a dedicated area, away from that of DNA extraction and PCR. Appropriate controls, including reagents, water, DNA extraction components, and bacterial DNA (which should not be amplified with the selected primers), should also be included routinely. Additionally, sequences can be aligned with a variety of taxa in order to eliminate those samples that clearly do not represent amplification of the target DNA.

Area 1: Reagent Preparation Area

This area should be supplied with a dead air box and UV light source within which all reagent mixtures can be prepared. All enzymes, reagents, and primers should be prepared for use and maintained in this room and not transported to other areas. No DNA, from any source, should be introduced into this area. This area must also contain a small freezer for storage of enzymes and reagents. All PCR setups must take place in this area. Lab coats, gloves, and masks are to be kept in here in order to be continually irradiated by UV light when not in use. Only authorized personnel should be allowed entry. The DNA must not be taken into this area at any time.

Area 2: Sample Preparation Area

Preparation of fossil tissues and extraction of DNA should be carried out in a laminar flow hood, on a work table or bench top that is continually bathed in UV light when not in use. This will denature any free DNA, making it unavailable for amplification. Reaction mixtures should be transported to this room where template DNA would be added. Reagents, gloves, lab coats, and pipettes for tissue prep should only be kept in this area and not transported to and from other areas.

Area 3: PCR and Gel Electrophoresis Area

Thermal cyclers and electrophoretic equipment should be located in this area. Resulting PCR products would be examined by gel electrophoresis, ethidium-bromide staining, and UV visualization for proper size. Products for reamplification would be cored out of the gel. Eventually, cloning and sequencing could be carried out in this area as well.

Tissue Handling and Processing

Tissues used for ancient DNA work must be handled at all times under aseptic conditions. Sterile instruments and containers must be used and gloves worn at all times when handling the fossils or instruments to be used with the fossils. Handling of sample must be done in areas where introduction of contaminant DNA is minimized through UV irradiation and frequent surface sterilizations with a 10% bleach solution.

EXTRACTION OF DNA FROM FOSSILS USING SILICA PARTICLES

The entire procedure must be performed in a previously decontaminated safety cabinet to avoid contamination with extraneous DNA in a room (or location) exclusively dedicated to the extraction of fossil DNA, and no extant DNA should ever be either extracted or handled in this area. The safety cabinets can be decontaminated by thoroughly washing with 10% bleach followed by two washes with 70% ethanol. Do not allow the 10% bleach to stay in contact with stainless steel surfaces for extended periods of time because it will eventually erode the surfaces. UV light should be left on in the safety cabinet while the cabinet is not in use. Additionally, the HEPAS filters should be replaced frequently to minimize the risk of microbial contamination of the samples. The DNA extractions should be performed by personnel wearing sterile gloves, masks, and head gear in order to avoid contamination of the samples with human DNA from hair, saliva, or skin cells.

Because the silica particles in the presence of chaotropic salts tend to indiscriminately bind DNA, regardless of source, it is imperative that at least two mock extractions be conducted in parallel to the fossil DNA extraction in order to detect the presence of contaminating DNA bound to the silica matrix. Mock extracts should also be subjected to the same PCR amplifications as the presumed fossil DNA samples. In addition, a reagent control (PCR reaction mixture with water rather than DNA) must be performed with every PCR assay.

PROTOCOL FOR PREPARATION OF DNA-BINDING SILICA MATRIX

REAGENTS REQUIRED

- silica matrix suspension (Sigma)
- extraction GuSCN buffer
- wash GuSCN buffer
- 70% ethanol
- acetone
- sterile TE buffer (0.1 M Tris, 0.1 M EDTA, pH 8.0)

PREPARATION OF REAGENTS (from Boom et al. [2])

Silica Matrix Suspension

(1) Suspend 30 g coarse SiO_2 (Sigma # S 5631) in 200 ml of filter-

sterilized, deionized water in a sterile, acid washed 250-ml graduated glass cylinder then bring the total volume to 250 ml. Cover the opening with sterile aluminum foil and allow the silica particles to sediment overnight at room temperature.

(2) Decant or suction-off 215 ml of the aqueous phase and resuspend the sediment, with vigorous shaking, in filter-sterilized, deionized water in a 250-ml graduated glass cylinder and then bring the total volume to 250 ml. Allow the resuspended silica particles to settle for five hours at room temperature.

(3) Remove 220 ml of the aqueous phase and add to the silica sediment 300 μl of 3.84 M HCl (32% weight/volume) to adjust the pH of the suspension to 2.0.

(4) Aliquot 5–10 ml the pH-adjusted silica particles in amber-colored, screw-capped tubes and autoclave at 121°C for twenty minutes. Store at room temperature in a dark, cool place until used. Store for ≤6 months.

Extraction GuSCN Buffer

(1) Dissolve 120 g of guanidinium thiocyanate (GuSCN) (Sigma) in 100 ml of filter-sterilized 0.1 M Tris-HCl (pH 6.4) by heating at 60°C with stirring until the GuSCN is completely dissolved.

(2) Add 22 ml of filter-sterilized 0.2 M EDTA (pH 8.0) and 2.6 g of Triton X-100 (Sigma # T 8787) and mix well.

(3) Store in the dark at room temperature. This solution is stable for at least four weeks when stored in an amber-colored bottle in a dark, cool place.

Wash GuSCN Buffer

(1) Dissolve 120 g of guanidinium thiocyanate (GuSCN) (Sigma) in 100 ml of filter-sterilized 0.1 M Tris-HCl (pH 6.4) by heating at 60°C with stirring until the GuSCN is completely dissolved.

(2) Store in the dark at room temperature. This solution is stable for at least four weeks when stored in an amber-colored bottle in a dark place.

PROCEDURE FOR EXTRACTING DNA FROM FOSSIL BONES
(from Höss and Pääbo [29])

(1) Remove 0.5–1.0 mm of external bone surface with a DREMEL Moto-Tool (DREMEN, Inc., Racine, WI), or similar grinding apparatus, to remove any contaminating DNA from the surface of the bones.

(2) Place 0.5–1.0 g of cortical or trabecular bone in a sterile mortar and cover with liquid nitrogen until the N_2 is completely evaporated. Grind the bone to a fine powder with a sterile pestle. Alternatively, the fossil bone can be made into a fine powder using a freezer mill (Spex Industries, Edison, NJ).

(3) Place approximately 0.5 g of powdered bone into a sterile, screw-capped, 15 ml centrifuge tube containing 1000 μl of extraction GuSCN buffer. Seal the tube well to avoid spillage, and incubate at 55–60°C for five to twelve hours. The bone mixture can be stored in an EnvironShaker dry bath (Eppendorf) or in a standard dry bath with occasional shaking.

(4) Centrifuge the bone suspension at 5,000 × g for five minutes at room temperature. Collect 500 μl of supernatant with a 1000 μl micropipet outfitted with an aerosol-resistant tip.

(5) Add the 500 μl of bone extract to a sterile, screw-capped, (with o-ring) 1.5-ml microcentrifuge tube containing 500 μl of extraction GuSCN buffer and 40 μl of the silica matrix suspension. Incubate at room temperature for ten to fifteen minutes. Centrifuge for thirty seconds at maximum speed in a microcentrifuge to pellet the silica particles. Discard the supernatant. The supernatant fluid may be reextracted as described in step 4 or stored at −20°C for protein/polypeptide extractions and analyses.

(6) Wash the pellet twice with 1 ml of wash GuSCN buffer, twice with 1 ml of 70% ethanol, and once with acetone. After discarding the acetone wash, dry the silica pellet in a heat block set at 55–60°C. Do not use a DryVac to evaporate the acetone because it may result in the contamination of the sample from aerosolized DNA from previous evaporations.

(7) Resuspend the dried silica pellet in 50 μl of filter-sterilized TE buffer (or deionized water) then incubate at 55–60°C for ten minutes. Centrifuge at maximum speed for thirty seconds and transfer the supernatant fluid (that contains the extracted DNA) to a sterile microcentrifuge tube. Repeat the DNA elution step with an additional 50 μl of eluent and pool the eluents. A third elution step is generally not necessary because ≥95% of the extracted DNA will be eluted during the first two elutions.

(8) Store the extracted DNA at −20°C until needed.

PROCEDURE FOR EXTRACTING DNA FROM MUSEUM SKINS

(1) Decontaminate both internal and external surfaces of the museum skin with a cotton swab soaked in 10% bleach. Remove the bleach with a

swab soaked in 70% ethanol (filter-sterilized) and a swab soaked in filter-sterilized, deionized water. Alternatively, bleach could be inactivated with 1% sterile sodium thiosulfate ($Na_2S_2O_3$).

(2) Place a 1–2 mm² piece of surface-decontaminated skin in a sterile mortar and cover with liquid nitrogen until the N_2 is completely evaporated. Grind the tissue to a fine powder with a sterile pestle.

(3) Place the powdered skin into a sterile, screw-capped, 15-ml centrifuge tube containing 500 μl of extraction GuSCN buffer. Seal the tube well to avoid spillage, and incubate at 55–60°C for five to twelve hours. The tissue mixture can be stored in an EnvironShaker dry bath (Eppendorf) or in a standard dry bath with occasional shaking. Follow steps 4–8 as described previously for bone extractions.

PROCEDURE FOR EXTRACTING DNA FROM AMBER INCLUSIONS

Decontamination of the Amber

(1) Immerse the amber piece with the desired inclusion (showing evidence of tissue preservations as determined by microscopic examination) in 25 ml of 2% glutaraldehyde in Tris-Cl (pH 7.2) and 0.1% Triton X-100 in a sterile, 50-ml, screw-capped centrifuge tube. Place the tube, with the lid on but looose, inside a vacuum oven set at 37°C, close the door tightly and incubate the amber piece under negative pressure overnight.

(2) Remove the amber piece from the glutaraldehyde solution with sterile forceps and rinse three times in 250 ml of sterile water.

(3) Immerse the amber piece in 25 ml of 10% bleach in a sterile, 50-ml, screw-capped centrifuge tube and place in a vacuum oven set at 37°C. Incubate the amber piece under negative pressure for two hours, then rinse three times as described in step 2 above.

(4) Immerse the amber piece in 25 ml of filter-sterilized 70% ethanol in a sterile, 50-ml, screw-capped centrifuge tube and place in a vacuum oven set at 37°C. Incubate the amber piece under negative pressure for two hours then ignite the alcohol to dry the piece.

Extraction of Tissues from Amber Inclusions

(1) Place the decontaminated amber piece in a high-walled dissecting dish that had been previously autoclaved at 121°C for one hour. Cover the amber piece with liquid nitrogen until all the N_2 evaporates. At this point the amber may exhibit cracks as it warms up. If not, a few drops of hot, sterile water may expedite the cracking.

(2) Expose the inclusion tissue by prying the amber segments apart with a pair of sterile, 27-gauge needles, attached to tuberculin syringes (Figure 3). This and subsequent steps may be carried out with the aid of a dissecting microscope that had been meticulously washed with 10% bleach, rinsed with sterile 1% $Na_2S_2O_3$, and stored under UV light for one hour prior to use.

(3) Remove the inclusion tissue with sterile needles and transfer to a sterile microcentrifuge tube containing 10 μl of sterile TE buffer. Mince the tissue apart with sterile needles until a fine suspension is attained.

Extraction of DNA from Amber Inclusions

An alternative procedure for the extraction of DNA using guanidinium thiocyanate and glass powder is described by Cano and Poinar [3].

(1) Place the minced tissue from step 3 above into a sterile, screw-capped (with an o-ring), 1.5-ml microcentrifuge tube containing 500 μl of extraction GuSCN buffer. Seal the tube well to avoid spillage, and incubate at 55–60°C for five to twelve hours. The bone mixture can be stored in an EnvironShaker dry bath (Eppendorf) or in a standard dry bath with occasional shaking. Follow steps 4–8 as described above for bone extractions. Because many amber inclusions weigh less than 1 mg, the extracted DNA should be eluted from the silica matrix with ≤ 1/2 the recommended volume for bone or museum skins.

EXTRACTING DNA FROM FOSSILS USING CHELEX-100

DNA extractions from extinct species can be carried out by the Chelex extraction method described by Walsh et al. [62]. It has been shown that this method can be used to extract amplifiable quantities of DNA rapidly and efficiently from most biological samples. This method is particularly well suited for the extraction of DNA from samples containing partially degraded DNA likely to exist in forensic materials and in tissues from amber inclusions [4,7]. Additionally, Chelex extractions of DNA can be performed using < 1 mg of sample, which is often all that is available in amber inclusions or museum skins.

This method has been used successfully in the laboratory to extract amplifiable quantities of DNA from *Proplebeia dominicana, Libanorhinus succinus,* and *Hymenaea protera* in 25–120 myo amber fossils as well as from herbarium specimens.

The entire procedure must be performed in a previously decontaminated safety cabinet to avoid contamination with extraneous DNA. This room (or location) is dedicated to the extraction of fossil DNA and no extant

Figure 3 Photomicrograph illustrating the extraction of abdominal tissue from an amber-entombed bee using 27-gauge needles.

DNA should be extracted or handled in this area. The safety cabinets can be decontaminated by thoroughly washing with 10% bleach followed by two washes with 70% ethanol. Ultraviolet light should be left on while the safety cabinet is not in use. The DNA extractions should be performed by personnel wearing sterile gloves, masks, and head gear in order to avoid contamination of the samples with human DNA from hair, saliva, or skin cells.

PROTOCOL FOR PREPARATION OF CHELEX-100 REAGENTS

MATERIALS REQUIRED

- 50-ml sterile, screw-capped centrifuge tube with 5-ml graduations
- filter-sterilized water
- Chelex-100 resin, 100–200 mesh (BioRad # 142-2832)

PREPARATION OF 5% (WEIGHT/VOLUME) CHELEX-100

(1) Weigh 1 g of Chelex-100 resin in a tared, 50-ml sterile, graduated, screw-capped centrifuge tube. Add filter-sterilized water to the 20-ml mark of the tube. If possible, these steps should be carried out in a safety cabinet decontaminated as described above.

(2) Tighten lid and mix well by vortexing. Store at 5°C overnight prior to use. Dispense in 1-ml aliquots using a micropipet outfitted with an aerosol-resistant tip, making sure that the stock suspension is well mixed. Store at 5°C in a box. Handle with gloved hands and take out only the number of tubes needed to carry out the extraction procedure. Each tube contains enough 5% Chelex-100 for one extraction and one control.

EXTRACTION OF DNA FROM FOSSILS USING CHELEX-100

(1) Prepare the tissue for extraction by mincing or grinding as described above for bone, museum skins, or amber inclusions.

(2) Resuspend the tissue in 500 µl of 5% Chelex-100 in sterile, screw-capped (with o-ring) microcentrifuge tube. Mix well and place in shaker incubator, water bath, or heat block at 55°C overnight. After the incubation period, mix well by vortexing for thirty seconds.

(3) Incubate at 95°C for five minutes. Mix well by vortexing for thirty seconds.

(4) Centrifuge at maximum speed for one minute to pellet the resin beads. Transfer 392 µl of the supernatant fluid (containing the eluted

DNA) to a sterile, microcentrifuge tube and add 8 μl of a 50× solution of filter-sterilized TE buffer.

(5) Store at $-20°C$ until needed.

THE POLYMERASE CHAIN REACTION ASSAY

Once the DNA from fossils has been successfully extracted, it is now ready for enzymatic amplification. Needless to say, gene selection and primer design are of prime importance and will depend on the goals of the amplification assay. Since each target DNA and its corresponding primer pair(s) are unique, the reaction, conditions, and thermal-cycling protocol will vary with each sequence; therefore, the assay must be optimized each time a new primer set is used (see Chapter 2). The subsequent paragraphs can only suggest reasonable starting points for the optimization of the assay. The use of the Stoffel fragment of Taq polymerase is recommended for initial studies of fossil DNA because this enzyme is more tolerant to fluctuations in Mg^{2+} concentrations, and therefore, would increase the chances for initial success.

Many fossil samples have tannins, porphyrins, hematin, and other inhibitors of the PCR reaction. For this reason the inclusion of bovine serum albumin (BSA fraction V, Sigma) in the reagent mixture at concentrations of 2 μg/ml to palliate the inhibitory activity of fossil DNA contaminants is recommended.

Also, to reduce spurious hybridization of primers to nonhomologous target DNA sequences, some modification of a "hot-start" PCR should be used. The following method has been largely successful in the laboratory and does not require the separate addition of polymerase to each tube or the use of wax beads. In essence, the reaction mixture and all the reagents are maintained on ice throughout the preparation, dispensing of the mixture into the tubes, and the addition of the template to the mixture. While this is done, a soak cycle of 80°C for five minutes is programmed into the cycler. When the heat block of the thermocycler reaches 80°C, the tubes are removed from the ice and placed immediately on the heat block. From then on, the thermal-cycling protocol proceeds normally.

PREPARATION OF THE MASTER MIXTURE

For most studies 25-μl reactions are all that is necessary; certainly volumes in excess of 100 μl per reaction are never warranted. Furthermore, the efficiency of heat transfer is greater in smaller volumes than larger ones. The protocol below is a representative example of an amplification using 25 μl of volume.

(1) Prepare a master mixture in Area #1 (Reagent Preparation Area) using

Table 3 as an example. Determine the total number of samples, including the controls, and add one more to allow for pipeting errors. To do this, label the tubes properly and place them, along with all the reagents, in an ice bath. Use hat, mask, coat, and gloves throughout the entire procedure.

(2) Still in Area 1, Aliquot 20 μl of the master mixture into each of the tubes and close lid. Still in ice, transport the tubes with the reagent mixture to Area 2 (Sample Preparation Area).

(3) Add 5 μl of template DNA, including a weak positive control and a negative control, to the appropriate tube. For each template, open the appropriate tube, add the template followed by 25 μl of sterile mineral oil (if necessary), and close the lid. Do not open the next tube containing the reagent mixture until the previous tube has been sealed. There should be only one tube open at any given time. All other tubes should be sealed.

(4) Transport the tubes, still in ice, to the thermal cycler. Turn on the thermal cycler and allow the heat block to reach a temperature of 80°C. When the heat block has reached the 80°C, diligently place the tubes in the heat block and close the lid. Table 4 outlines a typical temperature program utilized in the ancient DNA laboratory.

(5) Analyze the PCR product by agarose gel electrophoresis or other suitable methods.

SEQUENCING OF AMPLIFICATION PRODUCTS

There are many suitable protocols available for determining nucleotide sequencing of PCR products, both from clones or directly from PCR reac-

TABLE 3. Reagent Concentrations for a Typical PCR Assay.

Reagent	Stock Conc.	Volume in μl	Final Conc.	Number Samples	Total Volume
Template	—	5.0	5 fM		—
ddH$_2$O	—	8.9	—		89 μl
10× buffer[a]	10×	2.5	1×		25 μl
BSA	20 μg/ml	2.5	2 μg/ml		25 μl
MgCl$_2$	25 mM	2.0	2 mM	10	20 μl
dNTP	2.5 mM	2.0	0.2 mM		20 μl
Primer #1	10 μM	1.0	0.4 μM		10 μl
Primer #2	10 μM	1.0	0.4 μM		10 μl
Stoffel Taq	10 U/μl	0.1	1 U		1 μl

[a]The pH of the buffer is extremely important in order to preserve the integrity of the ancient DNA template. Tris-based buffers should have a room-temperature pH of 8.8 so that at higher temperatures the pH of the buffer is still not acidic and damaging to the DNA templates.
Total volume: 25 μl.
Number of tubes: 10.

TABLE 4. Temperature Cycling Profile in a Typical PCR Assay.

Program	Temperature	Time	Reps
1	80°C	5 min.	1
2	95°C	2 min.	1
3a	94°C	45 sec.	
3b	45°C	60 sec.	40
3c	72°C	30 sec.	
4	72°C	10 min.	1
5	5°C	hold	1

tions. These include single- and double-stranded template sequencing with Sequenase (USB, Cleveland, OH), cycle sequencing, and other techniques utilizing thermoresistant DNA polymerases. Each has its advantages and disadvantages, which must be evaluated by the investigator as best suited for the intended goal of the project.

It should be noted, however, that direct sequencing of PCR products normally yields a "consensus" sequence because the PCR product represents a "pool" of individual amplicons reflecting both template variation, template integrity, and polymerase fidelity. Sequences proceeding from cloned amplicons represent the sequence of that single amplicon ligated to the vector. It is not a consensus sequence and might reflect both template variations and/or polymerase errors. When sequencing cloned amplicons, it is recommended that a minimum of six different clones be used to generate a "consensus" sequence. Alternatively, purified plasmid DNA from ten to twenty clones may be pooled and a single sequencing reaction conducted because this represents a "consensus" sequence of the ten to twenty clones pooled.

Regardless of the method used and the approach to sequencing, sequence reproducibility and comparison to those of known, related taxa should be performed to increase the degree of reliability on the sequence data generated from ancient DNA.

REFERENCES

1 Adegoke, J. A., B. O. Ighavini, and R. O. Onuigbo. 1991. Characteristic features of the sonicated DNA of *Agama agama agama* L. (Reptilia, Agamidae) on hydroxyapatite columns, using mouse DNA as a reference. *Genetica.* 83:171–180.

2 Boom, R., C. J. A. Sol, M. M. M. Salimans, C. L. Jansen, P. M. E. Wertheim-van Dillen, and J. Vander Noordaa. 1990. Rapid and simple method for purification of nucleic acids. *J. Clin. Microbiol.* 28:495–503.

3 Cano, R. J. and H. N. Poinar. 1993. Rapid isolation of DNA from fossil and museum specimens. *BioTechniques.* 15(3):8–11.

4 Cano, R. J., H. N. Poinar, D. W. Roubik, and G. O. Poinar, Jr. 1992. Enzymatic amplification and nucleotide sequencing of portions of the 18s rRNA gene of the bee *Proplebeia dominicana* (Apidae: Hymenoptera) isolated from 25–40 million year old Dominican amber. *Med. Sci. Res.* 20:619–623.

5 Cano, R. J., M. Borucki, M. Higby-Schweitzer, H. Poinar, G. O. Poinar, Jr., and K. Pollard. 1994. *Bacillus* DNA in amber: a window to an ancient symbiosis? *Appl. Environ. Microbiol.* 60:2164–2167.

6 Cano, R. J., H. Poinar, and G. O. Poinar, Jr. 1992. Isolation and partial characterization of DNA from the bee *Proplebeia dominicana* (Apidae:Hymenoptera) in 25–40 million year old amber. *Med. Sci. Res.* 20:249–251.

7 Cano, R. J., H. Poinar, N. Pieniazek, A. Acra, and G. O. Poinar, Jr. 1993. Amplification and sequencing of DNA from a 120–135 million year old weevil. *Nature.* 363:536–538.

8 Chinsamy, A. 1991. Physiological Implications of the Bone Histology of *Syntarsus rhodesiensis* (Saurischia: Theropoda). *Palaeontol. Africana.* 27:77–82.

9 Collins, M. J., G. Muyzer, P. Westbroek, G. B. Curry, P. A. Sandberg, S. J. Xu, R. Quinn and D. McKinnon, 1991. Preservation of fossil biopolymeric structures: Conclusive immunological evidence. *Geochim. Cosmochim. Acta,* 55: 2253–2257.

10 de Jong, E. W., P. Westbroek, J. F. Westbroek, and J. W. Bruning, 1974. Preservation of antigenic properties of macromolecules over 70 Myr. *Nature.* 252:63–64.

11 de Ricqles, A. 1983. Cyclical growth in the long limb bones of a sauropod dinosaur. *Acta Palaeontol. Pol.* 28(1–2):225–232.

12 de Ricqles, A. 1980. Tissue Structures of Dinosaur Bone; Functional Significance and Possible Relation to Dinosaur Physiology. In: *A Cold Look at the Warm Blooded Dinosaurs,* Roger D. K. Thomas and Everett C. Olson, eds. pp. 103–139.

13 deQueiroz, K. 1985. The ontogenetic method for determining character state polarity and its relevance to phylogenetic systematics. *Sys. Zool.* 34:280–299.

14 DeSalle, R., J. Gatesy, W. Wheeler, and D. Grimaldi. 1992. DNA sequences from a fossil termite in oligo-miocene amber and their phylogenetic relationships. *Science.* 257:5078–5081.

15 Donoghue, M. J., J. A. Doyle, J. Gauthier, A. G. Kluge, and T. Rowe. 1989. The importance of fossils in phylogeny reconstruction. *Ann. Rev. of Ecol. Sys.* 20:431–460.

16 Eernisse, D. and A. G. Kluge. 1993. Taxonomic congruence versus total evidence, and amniote phylogeny inferred from fossils, molecules, and morphology. *Mol. Biol. Evol.* 10(6):1170–1195.

17 Eglinton, Geoffrey and G. A. Logan. 1991. Molecular preservation. *Phil. Trans. R. Soc. Lond. B.* 333:315–328.

18 Felsenstein, J. 1989. PHYLIP—Phylogeny inference package (Version 3.2). *Cladistics.* 5:164–166.

19 Golenberg, E. M., D. E. Giannasi, M. T. Glegg, C. J. Smiley, M. Durbin, D. Henderson, and G. Zurawski. 1990. Chloroplast DNA sequence from Miocene *Magnolia* species. *Nature.* 344:656–658.

20 Grossman, Lawrence. 1991. Repair of damaged DNA. *Encyclopedia of Human Biology, Vol. 6,* pp. 547–553.

21 Gurley, L. R., J. G. Valdez, W. D. Spall, B. F. Smith, and D. D. Gillette, 1991. Proteins in the fossil bone of the dinosaur *Seismosaurus. J. Prot. Chem.* 10(1):75–90.

22 Hagelberg, E. and J. B. Clegg, 1993. Genetic polymorphisms in prehistoric Pacific islanders determined by analysis of ancient bone DNA. *Proc. R. Soc. Lond. Biol.* 252(1334):163–170.

23 Hagelberg, E. and J. B. Clegg. 1991. Isolation and characterization of DNA from archaeological bone. *Proc. R. Soc. Lond. Biol.* 244:45–50.

24 Handt, O., M. Höss, M. Krings, and S. Pääbo. 1993. Ancient DNA: methodological challenges. *Experientia.* 50:524–529.

25 Higgins, D. G. 1992. Sequence ordinators: a multivariate analysis approach to analysing large sequences data sets. *Comput. Appl. Biosci.* 8:15–21.

26 Higgins, D. G., A. J. Bleasby, and R. Fuchs. 1992. CLUSTAL V: Improved software for multiple sequence alignment. *Comput. Appl. Biosci.* 8:189–191.

27 Higuchi, R. and A. C. Wilson, 1984. Recovery of DNA from extinct species. *Fed. Proc.* 43:1557.

28 Higuchi, R., B. Bowman, M. Freiberger, O. A. Ryder, and A. C. Wilson. 1984. DNA sequences from the Quagga, an extinct member of the horse family. *Nature.* 312:282–284.

29 Höss, M. and S. Pääbo. 1993. Silica based method of DNA extraction from ancient bone. *Nucleic Acid Res.* 21:3913–3914.

30 Johnson, P. H., C. B. Olson, and M. Goodman, 1985. Isolation and characterization of the deoxyribonucleic acid from tissue of the woolly mammoth, *Mammuthus primigenius. Comp. Biochem. Physio. [B].* 81:1045–51.

31 Kocher, T. D., W. K. Thomas, A. Meyer, S. V. Edwards, S. Pääbo, F. X. Villablanca, and A. C. Wilson. 1989. Dynamics of mitochondrial DNA evolution in animals: Amplification and sequencing with conserved primers. *Proc. Natl. Acad. Sci., USA.* 86:6196–6200.

32 Koop, B. F., D. A. Tagle, M. Goodman, and J. L. Slightom. 1989. A molecular view of primate phylogeny and important systematic and evolutionary questions. *Mol. Biol. Evol.* 6:580–612.

33 Kumar, S., K. Tamura, and M. Nei. 1993. *MEGA: Molecular Evolutionary Genetic Analysis,* Version 1.0. Pennsylvania State University, University Park, PA.

34 Langenheim, J. H. 1990. Plant resins. *Am. Sci.* 78:16–28.

35 Lindahl, T. 1993. Instability and decay of the primary structure of DNA. *Nature.* 362:709–715.

36 Miller, P. L., P. M. Nadkarni, and N. M. Carriero. 1991. Parallel computation and FASTA. Confronting the problem of parallel database search for a fast sequence comparison algorithm. *Comp. Appl. Biol. Sci.* 7:71–78.

37 Moran, N., M. A. Munson, P. Baumann, and H. Ishikawa. 1993. A molecular clock in endosymbiotic bacteria is calibrated using the insect host. *Proc. R. Soc. Lond. Biol.* 253:167–171.

38 Mullis, K., F. Faloona, S. Scharf, R. Saiki, G. Horn, and H. Erlich. 1986. Specific enzymatic amplification of DNA in vitro. The polymerase chain reaction. *Cold Spring Harb. Symp. Quant. Biol.* Pt 1:263–273.

39 Muyzer, G. and P. Westbroek. 1989. An immunohistochemical technique for the localization of preserved biopolymeric remains in fossils. *Geochim. Cosmochim. Acta.* 53:1699–1702.

40 Muyzer, G., P. Sandberg, M. H. J. Knapen, C. Vermeer, M. Collins, and P. Westbroek. 1992. Preservation of the bone protein osteocalcin in dinosaurs. *Geology.* 20:871–874.

41 Naito, E., K. Dewa, H. Ymanouchi, and R. Kominami, 1991. Ribosomal Ribonucleic Acid (rRNA) Gene Typing for Species Identification. *J. Forensic Sci.* 37(2):396–403.

42 Ochman, H. and A. W. Wilson. 1987. Evolution in bacteria: evidence for a universal substitution rate in cellular genomes. *J. Mol. Evol.* 26:74–86.

43 Pääbo, S. 1985. Molecular cloning of ancient Egyptian mummy DNA. *Nature.* 314:644–645.

44 Pääbo, S. 1989. Ancient DNA: extraction, characterization, molecular cloning, and enzymatic amplification. *Proc. the Natl. Acad. Sci. USA.* 86:1939–1943.

45 Pääbo, S., R. G. Higuchi, and A. C. Wilson. 1989. Ancient DNA and the polymerase chain reaction. *J. Biol. Chem.* 264:9709–9712.

46 Persing, D. H., S. R. Telford, P. N. Rys, D. E. Dodge, T. J. White, S. E. Malawista, and A. Spielman. 1993. Detection of *Borrelia burgdorferi* DNA in museum specimens of *Ixodes dammini* ticks. *Science.* 249:1420–1423.

47 Poinar, H., G. O. Poinar, Jr., and R. J. Cano. 1993. Molecular phylogeny of an extinct legume *(Hymenaea protera)* from Dominican amber. *Nature.* 363:677.

48 Poinar, Jr., G. O. and R. Hess. 1985. Ultrastructure of 40 million year old insect tissue. *Science.* 215:1241–1242.

49 Poinar, Jr., G. O. 1993. The range of life in amber: significance and implications in DNA studies. *Experientia.* 50:536–542.

50 Poinar, Jr., G. O. and R. Hess. 1985. Preservative qualities of recent and fossil resins: electron micrograph studies on the tissue preserved in baltic amber. *J. Baltic Stud.* 16(3):222–230.

51 Reid, R. E. H. 1984a. Primary bone and dinosaurian physiology. *Geological Magazine.* 121(6):589–598.

52 Reid, R. E. H. 1984. The histology of dinosaur bone, and its possible bearing on dinosaur physiology. *Symp. Zool. Soc. Lond.* 52:629–663.

53 Reid, R. E. H. 1985. On supposed haversian bone from the hadrosaur anatosaurus, and the nature of compact bone in dinosaurs. *J. Paleontol.* 59:140–148.

54 Romanowski, G., M. G. Lorenz, and W. Wackernagel. 1991. Adsorption of plasmid DNA to mineral surfaces and protection against DNase I. *Appl. Environ. Microbiol.* 57:1047–1061.

55 Sanger, F., S. Nicklen, and A. R. Coulson. 1977. DNA sequencing with chain-terminating inhibitors. *Proc. Natl. Acad. Sci. USA.* 74:5463–5466.

56 Simon, C., F. Frati, A. Beckenbach, B. Crespi, H. Liu, and P. Flook. 1994. Evolution, weighing, and phylogenetic utility of mitochondrial gene sequences and a compilation of conserved polymerase chain reaction primers. *Ann. Entomol. Soc. Am.* 87:651–701.

57 Swofford, D. L. 1990. *Phylogenetic Analysis Using Parsimony,* Version 3.0, computer program, Illinois Natural History Survey, Champaign, IL.

58 Thomas, R. H., W. Schaffner, A. C. Wilson, and S. Pääbo. 1989. DNA phylogeny of the extinct marsupial wolf. *Nature.* 340:465–467.

59 Thomas, W. K., S. Pääbo, F. X. Villablanca, and A. C. Wilson. 1990. Spatial and temporal continuity of kangaroo rat populations shown by sequencing mitochondrial DNA from museum specimens. *J. Mol. Evol.* 31:101–112.

60 Van de Peer, Y., J. Neefs, and R. deWachter. 1990. Small ribosomal subunit RNA sequences, evolutionary relationships among different life forms, and mitochondrial origins. *J. Mol. Evol.* 30:463–476.

61 Varrichio, D. 1993. Bone microstructure of the upper Cretaceous theropod dinosaur *Troodon formosus*. *J. Vert. Paleontol.* 113:99–104.

62 Walsh, P. S., D. A. Metzger, and R. Higuchi. 1991. Chelex-100 as a medium for simple extraction of DNA for PCR-based typing from forensic material. *BioTechniques.* 10:506–513.

63 Wert, C. A. and M. Miller. 1988. The polymeric nature of amber. *Bull. Amer. Physiol. Soc.* 33:497.

64 White, T. J., T. Bruns, S. Lee and J. Taylor. 1990. Amplification and direct sequencing of fungal ribosomal RNA genes for phylogenetics. In: *PCR Protocols: A Guide to Methods and Applications,* Innis, M. A., D. H. Gelfand, J. S. Sninsky, and T. J. White (eds.) Academic Press, Inc., San Diego, CA, pp. 315–324.

Index